Subartu XXVIII

Holocene landscapes through time in the Fertile Crescent

Subartu

Subartu is edited by the European Centre for Upper Mesopotamian Studies

Manuscripts are to be submitted to:
Marc Lebeau, ECUMS-Brussels, 41 Boulevard A. Reyers, Bte 6, B-1030 Brussels, Belgium

Order forms to be mailed to:
Brepols Publishers, Begijnhof 67, B-2300 Turnhout, Belgium

Katleen DECKERS
(editor)

Holocene landscapes through time in the Fertile Crescent

BREPOLS

Katleen DECKERS (ed.)
Holocene landscapes through time in the Fertile Crescent
(=Subartu XXVIII), Brepols, Turnhout, 2011
A4, sewn, XII + 172 pages
Contents: Environmental archaeology, Anthracology, Archaeobotany, Geoarchaeology, OSL-dating
Area : Near East, Syria, Turkey

ISBN 978-2-503-54106-8
D/2011/0095/78

Table of contents

Acknowledgments

This volume was made possible thanks to the support and inspiration of many people, without whom this collaborative work would have never been realized.

First of all, I would like to thank the Belgische Stichting Roeping, for their valuable and flexible support of my vocation in researching environmental archaeology in the Near East. Additionally, this project prospered favourably from the "elite support" program of the Landesstiftung Baden-Württemberg. Some parts of the project were also made executable thanks to research conducted in the frame of the DFG-projects RI1193/4-1 and RI1193/4-2, under the direction of Dr. Simone Riehl. Final stages of this volume were realised thanks to a Margarete von Wrangell-Habilitationsstipendium, which is supported by the European Social Fonds in Baden-Württemberg.

Geoarchaeological fieldwork was possible thanks to several persons and projects. Firstly, in 2002 I was able to join the Tell Mozan project under the direction of Prof. Peter Pfälzner, Prof. Giorgio Buccellati, and Prof. Marilyn Kelly-Buccellati. This afforded me the opportunity to launch a geoarchaeological survey in the Upper Khabur. In 2003, this fieldwork was continued as part of the Kelly-Buccellati Mission in Mozan. I would like to thank them for their hospitality and support of my research interests.

In the spring of 2005, Prof. Harvey Weiss was kind enough to invite me to join the Tell Leilan excavations, where I was able to investigate the riverine history in the Leilan surroundings. I would like to thank him for his straightforward, uncomplicated, and very efficient support.

By summer of the same year, I was able to join the Tell Brak survey team and work with Prof. Tony Wilkinson, Prof. Henry Wright, and Prof. Joan Oates. I am very grateful for this collaboration because I learned a lot during the short time we shared in the field together. Fieldwork in the Brak surroundings was possible thanks to Prof. Joan Oates and Prof. Henry Wright. Additionally, we are also grateful for the support of Prof. Geoff Emberling and Prof. Helen McDonald.

Also in the summer of 2005, I went back to Mozan where I was fortunate enough to conduct a more thorough investigation of the palaeostream that ran through Mozan. Here, again, I experienced the strong support of Prof. Giorgio Buccellati and Prof. Marilyn Kelly-Buccellati and would like to express gratitude to them for their valuable help in making this possible. Furthermore, Prof. Marilyn Kelly-Buccellati greatly aided me with her expertise on pottery sherds from the Mozan geoarchaeological profile.

I am indebted to the General Directorate of Antiquities and Museums of Syria who kindly permitted the fieldwork. Without their support, the work contained in this volume would have been unattainable.

As for the charcoals, I am appreciative at getting the chance to collaborate with several excavation projects. Prof. Nicholas J. Conard invited me to investigate the Baaz and Kaus Kozah material. I am grateful for this opportunity and for the uncomplicated and fast exchange of data. Dr. Andrew Kandel, from the same team, delivered the necessary information at an exceptional rate and also provided some of the figures for chapter 1.

Many thanks are due to Prof. Peter Pfälzner for allowing me to examine the Qatna charcoals. I also wish to express my appreciation to Dr. Conrad Schmidt of Pfälzner's team for his continuous and efficient help in expanding my understanding of the contexts of the samples.

I was also fortunate enough to partner with several archaeological projects on the charcoal material mentioned in chapter 3. *Thank yous* are specifically extended to Prof. Peter Pfälzner for allowing me to evaluate the Tell Mozan charcoal remains, Dr. Uwe Finkbeiner for the Emar material, Prof. Edgar Peltenburg for the Jerablus charcoals, and Prof. Harvey Weiss for the Leilan charcoals.

In the summer of 2007, Dr. Hugues Pessin, who was at that time a Marie Curie fellow at the University of Sheffield, came over to Tübingen and we put together our charcoal data to gain an overview of the vegetation evolution in northern Syria/southern Turkey (see chapter 3). Dr. Hugues Pessin would like to thank Prof. Catherine Marro, Prof. Christine Kepinski, and Prof. Luc Bachelot that allowed him to investigate samples from respectively Horum Höyük, Tilbeshar, and Tell Shiukh Fawqani.

Dr. Dimitri Vandenberghe would like to acknowledge gratefully the technical assistance by Gilles Velghe and Jochem Temmerman for the dating. His luminescence research at Ghent University is financially supported by the Research Foundation Flanders (DV: Postdoctoral Fellow).

Many students and collaborators have been involved in this project and made valuable contributions to this volume. More precisely, Andrea Orendi, Nina Roser, Kerstin Schiele, Kirsten Eppler, Stefan Ertmer, and Christoph Wissing sorted botanical remains for many hours, days, and months. Christoph Wissing also undertook sedimentological analysis. Gang Wang worked for over more than a year on the GIS database, receiving additional assistance from Marina Monz. Camilla Reece-Trapp helped with English editing at the start of putting together this volume. Last but not least, M. Deva Jebb undertook the major part of editing this volume and also made many of its figures and tables. I would like to thank them all for their great work and enthusiasm!

Katleen Deckers

Introduction

This Subartu volume includes several articles that focus on landscape and environmental reconstruction in Syria and southern Turkey, and furthers our understanding of human-environment interactions in this area. A variety of methods are employed here, in order to gain insight into former environments and societies.

The volume starts with a chapter by Deckers and Conard (chapter 1) on the Neolithic transition, and presents results on the anthracological research at the sites of Baaz and Kaus Kozah. Aims of this paper were to reconstruct environmental conditions at the Pleistocene-Holocene border in the Damascus Province of Syria and to gain insight into the distribution of wild cereals. The authors of this chapter found that although *Amygdalus* steppe occurred during the Younger Dryas in this region, the samples that were associated with PPNA finds suggest that shortly after the Younger Dryas climatic conditions were moister. Furthermore, during the Younger Dryas this region seemed somewhat further away from the zone of dense stands of wild cereals, while shortly after, in the early PPNA, this region was probably located within this zone.

In the second chapter, by Deckers and Drechsler, we move to the Upper Khabur area of northeastern Syria and focus on the third millennium BC. The chapter provides an introduction to the settlement history of this area and investigates the land-use patterns, resource exploitation, and what environmental factors had an influence on the observed settlement distribution. Our analysis indicates that during the Bronze Age water, especially streams, but also soil moisture were important factors driving site selection in the Upper Khabur. The data for the Upper Khabur zone differ somewhat from the eastern Jezira, in that a three-tiered settlement system is not evident within the surroundings of every major centre. Our resource exploitation data further indicated that some larger centres clearly had dependent satellite villages, since they were located within the resource area of these centres. However, along the Wadi Avedji or Wadi Khanzir no overlap of resource exploitation zones was visible, which may indicate a degree of self-sufficiency in the villages. The density of occupation along the Wadi Jaghjagh and lack of sufficient cultivatable land within the surroundings of these sites suggests that this river was an important trade route. Finally, resource calculations that incorporate a tribute that is as high as the necessary harvest to feed the people indicate that resource limits were reached in many areas. Tribute, therefore, may have played a role in the observed settlement collapse at the end of the third millennium BC.

Chapter 3, by Deckers and Pessin, provides an overview of anthracological research in northern Syria/southern Turkey in order to gain insight into the relative impact by humans and climate on the environment in the third through second millennium BC. The Bronze Age vegetation in the area under investigation was generally much richer than the one observed today. In addition, the riverine vegetation was remarkably more diverse, and oak park woodland could still be found in northern Syria; analogous to this, *Pistacia* (woodland) steppe occurred as far south as Emar. Hence, the present data suggests that massive degradation, including deforestation, occurred within the research area. There are some hints both for human and climatic impact on the vegetation during the Bronze Age but, still, by the end of this period, the region was even more covered in woody vegetation than is seen today.

In chapter 4, by Deckers and Pustovoytov, we zoom in to a site of the Upper Khabur Basin, Tell Mozan, and investigate fluvial changes that took place in its surroundings. This chapter also includes geoarchaeological research on a ditch profile surrounding the site., Tell Mozan was located along an intermittent stream, even though today little is visible of this stream. Additionally, this chapter highlights that the inactive regular drainage pattern southeast of Mozan was not an irrigation system, but represents a natural palaeostream with a regular pattern. Finally, another revelation culled from this research indicates the ditch around the Bronze Age settlement did not permanently contain water. The infilling of the ditch started at around 2800 calibrated BC, or prior to that, and was especially intensive sometime between 2800 and 1000 BC. It may have been a sedimentological response to agricultural exploitation of the site's surroundings; for example, during the urban explosion of the Early Bronze Age or in the early Middle Bronze Age when increased human impact on the woody vegetation took place.

Chapter 5, by Goldberg, focuses on the micromorphological results from the ditch profile surrounding Mozan and contains some of the primary data on which chapter 4's interpretations are built.

In chapter 6, Deckers explores the riverine landscape changes that took place surrounding another site of the Upper Khabur Basin, more precisely the site of Tell Leilan. For the Leilan region, one of the most interesting details uncovered (chapter 6) is the canalisation of the Wadi Siblah, which was possibly made in relation to the construction of the lower town wall, circa 2600 BC. However, it may have also been a later feature, at the latest – roughly 1700 BC, during the final intense occupation of Leilan. The town's associated heightened need for water may relate to the large settlement expansion that took place circa 2600 BC.

In chapter 7, also by Deckers, geoarchaeological research focussed on the Tell Hamidi surroundings in the Upper Khabur Basin. Geoarchaeological research there has shown that the Wadi Jaghjagh had considerable palaeodischarge in the mid fourth millennium BC, the fifth century BC or somewhat later, and in the fifth century AD or somewhat later. The palaeochannel near Hamidi was meandering in the fifth century BC, although it may have been anastomosing by the mid fourth millennium BC. There is also evidence for aggradation from the third until the second millennium BC, which may relate to human impact on the landscape. Finally, the Jaghjagh was originally accompanied by a riverine gallery forest and *Phragmites* and *Cyperus* swamps. Today, these are lacking and were already scarce in the early 20th century AD.

The chapter that follows, by Riehl, provides details on the ostracod, fruit, and seed remains from a sedimentological profile discussed in chapter 7.

In chapter 9, by Wilkinson and Deckers, a geoarchaeological study addresses the loess profile at the site of Brak. The sedimentary sequence found at Brak indicates the accumulation of eolian loess was relatively steady and continuous between around 1500 BC and the 1st century BC. The dust-fall was probably related to the increased levels of eolian dust kicked up by local agricultural activity, as well as settlement on and around the tell. Nevertheless, there is no evidence for increased dust-fall at the end of the third millennium BC. The only evidence for changing rates of sedimentation was in the period between approximately the late second millennium BC and the 1st century BC, when there was a slight decrease in sedimentation and a palaeosol could develop. Furthermore, this may have corresponded with decreased occupation.

The basics and methods needed for Optically Stimulated Luminescence (OSL) dating of sediments is provided by Vandenberghe in chapter 10. He then continues with applications and evaluations of samples from the geomorphological fieldwork in the Upper Khabur Basin.

Vandenberghe's discussion is continued by Hilgers in chapter 11, where she reports on other OSL samples from the geomorphological fieldwork.

Subsequently, in chapter 12 by Deckers, charcoal-to-seed ratio investigations are undertaken to gain insight into what degree dung was used as fuel at two sites in Syria. Results indicate dung was used as fuel more often at sites in park woodlands (e.g. Tell Mozan) than at sites in riverine forests (e.g. Emar). This seems to suggest that the park woodland was relatively open compared to the riverine forest. However, it cannot be discounted that people used dung by preference, for example, for specific purposes. On the other hand, no special correlation was discovered between the use of dung as fuel and certain types of ovens or hearths.

Finally, chapter 13 focuses on anthracological research at Qatna, to understand the use of wood in the palace. Deckers discusses evidence found at Qatna Palace, in western Syria, which suggests kings had a larger impact on the vegetation than commoners. One such example calculated from the collected data revealed that to create the palace's roof, approximately 62.5 hectares of *Cedrus* forest would have been exploited.

With its array of information on the interplay between environment and human society, the editor hopes *Holocene landscapes through time in the Fertile Crescent* will both inspire and aid the research of those students and professionals in the field of Near Eastern (environmental) archaeology.

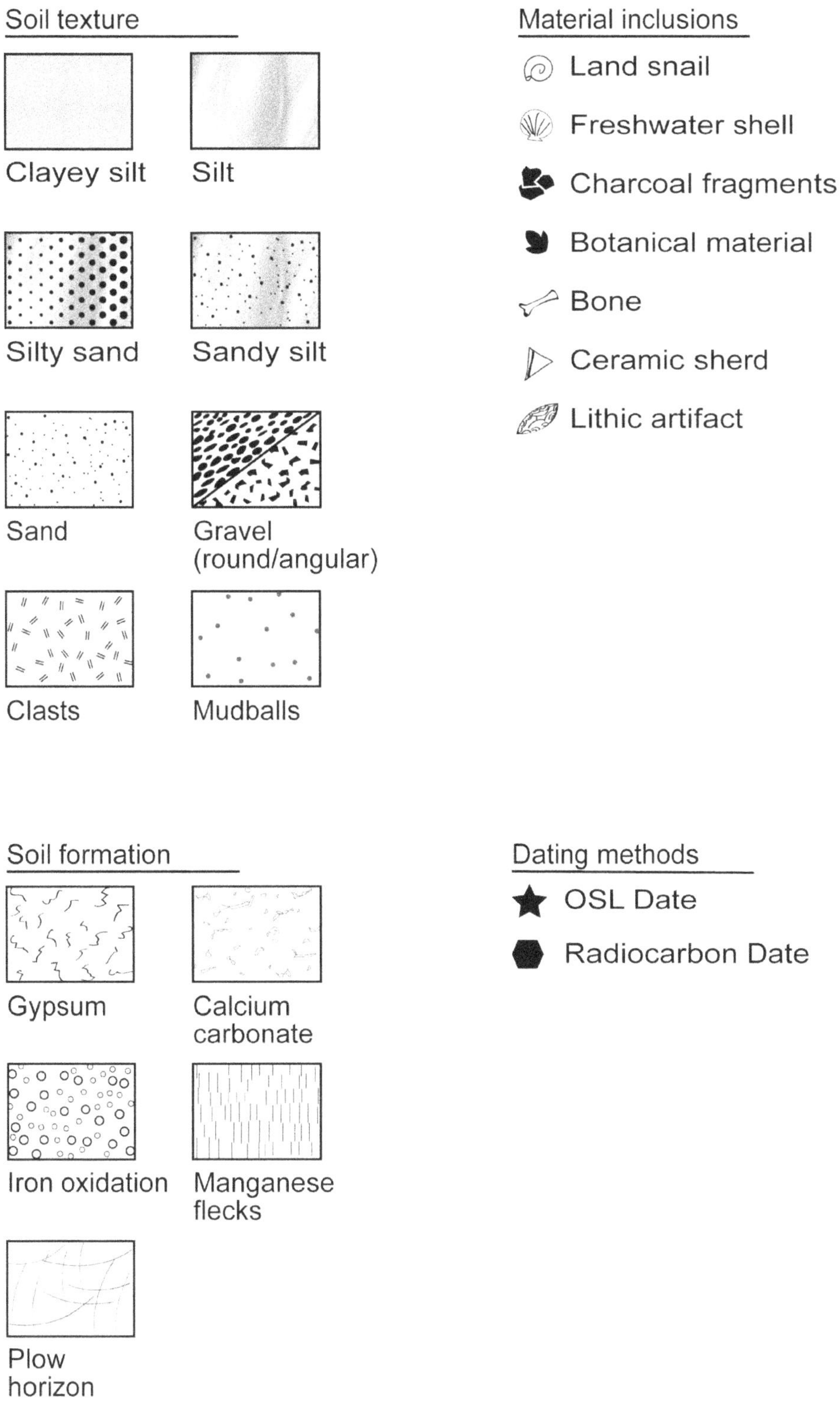
Soil texture
Clayey silt
Silt
Silty sand
Sandy silt
Sand
Gravel (round/angular)
Clasts
Mudballs
Material inclusions
Land snail
Freshwater shell
Charcoal fragments
Botanical material
Bone
Ceramic sherd
Lithic artifact
Soil formation
Gypsum
Calcium carbonate
Iron oxidation
Manganese flecks
Plow horizon
Dating methods
OSL Date
Radiocarbon Date

1. Vegetation and wood-use from the Pleistocene to the Holocene in the foothills of the eastern Anti-Lebanon

Katleen Deckers and Nicholas J. Conard

1. Introduction

At the end of the Pleistocene important changes took place in human subsistence. More precisely, the transition from hunting and gathering to farming and herding occurred. It remains difficult to clearly identify the earliest evidence of pre-domestic agriculture. While some place it in the Pre-Pottery Neolithic A (PPNA),[1] others find that clear indications only occur from the Pre-Pottery Neolithic B (PPNB) onwards.[2] However, it is generally agreed that this new subsistence mode was already developing prior to this point in time, as suggested by the increase in sedentism, storage capacities, and the dependence on wild cereals during the late Pleistocene.

These subsistence changes broadly coincided with environmental changes that marked the close of the Pleistocene, including: a worldwide increase in temperature that melted ice sheets and caused sea levels to rise, alterations in atmospheric circulation systems, and shifts in botanical and faunal zones. Thus many, although not all, investigators have turned to climatic and environmental change as a possible explanation for, or contributor to changes in subsistence strategies. Some of these models suggest that environmental scarcity related to dry climatic conditions of the Younger Dryas had an impact on the transition to agriculture,[3] while others stress that environmental wealth was the cause.[4]

Whatever the causes, a precondition for their intensive use and domestication is the availability of wild progenitors. While present day distributions of wild cereals are helpful for defining areas where agriculture may have originated, we must take into account that at the end of the Pleistocene dislocations in the distribution of vegetation occurred, including those of dense stands of wild cereals. The broad vegetation changes that took place from the Pleistocene to the Holocene have been documented through palynological research. Nevertheless, there have been large recurring chronological problems that until now could not be sufficiently clarified. For example, in the Levant, the Ghab chronology remains disputed and unresolved,[5] and problems persist with Huleh.[6] Additionally, there are large regions where no pollen archives could be obtained due to bad preservation conditions, leaving microregional developments in most cases indiscernible. We see this in the Damascus region, where Bottema[7] mentioned problems in gaining sufficient amounts of pollen from the lake basins of Utaibeh and Hejaneh. While Hussein[8] was able to retrieve better pollen samples from this same area, he was still unable to achieve a numerical chronology for the sequences. Some pollen data exists from archaeological sites within the Damascus region,[9] however, they are generally difficult to interpret in terms of vegetation reconstruction due to their low number of pollen counts and the possible influx of pollen from distant locales.

Dense stands of cereals typically occur in open areas within oak-Rosaceae (rose family) park-woodland, but especially occur on the border of this zone with the pistachio-almond woodland-steppe.[10] Due to the fact that Rosaceae species (to which almond also belongs), pistachio, and cereals are generally underrepresented in pollen diagrams, exact locations with dense stands of cereals could be overlooked when using palynological evidence alone. It is therefore obvious that anthracology should also play a key

---

1 Colledge 2002.

2 see e.g. Nesbitt 2002.

3 Bar-Yosef – Belfer-Cohen 1989; 1991; 2002; Bar-Yosef 1996; 2001; 2002.

4 Uerpmann 1979; Bottema 2002; Munro 2003.

5 e.g. Hillman 1996; Cappers *et al.* 2002.

6 see e.g. discussion in Cappers *et al.* 2002; Meadows 2005.

7 Bottema 1975-1977, 259.

8 Hussein 2006.

9 Kaiser *et al.* 1973; Deckers *et al.* 2009.

10 Hillman 1996.

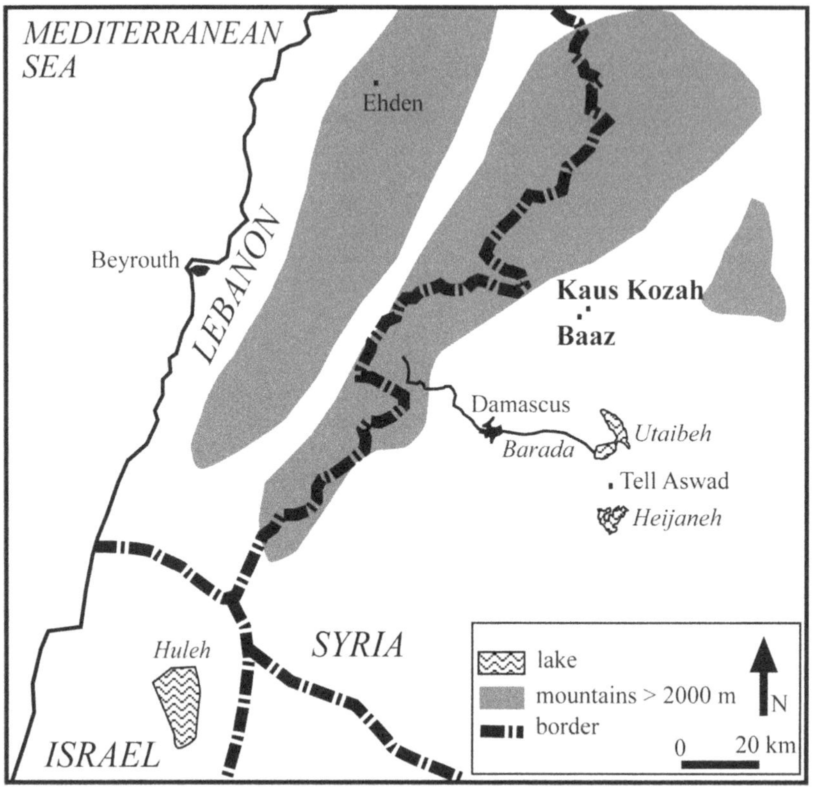

Fig. 1: Map showing the location of Baaz and Kaus Kozah in Syria.

role in reconstructing local vegetation conditions at archaeological sites. This is not only because of the better chronological resolution anthracology often provides compared to those of lake sediments, but also because of its possibility to recognise vegetation that is poorly visible in pollen diagrams.

The use of anthracology as a method to reconstruct the local vegetation is mainly based on the theory of "least effort," meaning firewood collection in the past most likely took place in woodland catchments situated closest to the settlement and species were probably collected in proportion to their occurrence in the site surroundings. Therefore, the frequencies and percentages of taxa are an indicator of the vegetation at the time of the human habitation. Of course, we must also take into account other factors distorting the vegetation reconstruction, like the fact that some taxa produce more deadwood than others and that these may be overrepresented in the samples; also, cultural preference, taphonomic processes, and retrieval biases may be present.[11]

In this chapter, charcoal remains from two archaeological sites with occupations within the Epipalaeolithic and Early Neolithic, Baaz Rockshelter and Kaus Kozah Cave, will be investigated in order to gain a better understanding of the vegetation at the end of the Pleistocene/start of the Holocene. We also examined the possible implications for human occupation and subsistence in the region.

2. The sites

The sites of Baaz and Kaus Kozah are located about three kilometres from one another and approximately thirty-five kilometres northeast of Damascus within a region that today receives an annual rainfall of about 200 mm (fig. 1). Both sites are located in the foothills of the Anti-Lebanon Mountains. The present day vegetation consists of steppe and is dominated by thorny bushes. Currently, few trees grow away from the local water sources or recently planted trees (almond orchards).

Baaz Rockshelter (figs. 2a, 2b, and 2c) overlooks the Jaba´deen Pass, which serves both as a reliable source of water due to its perennial springs and as a means for moving between the lowlands and the highlands of the westernmost Palmyride Mountains.[12] Kaus Kozah Cave (figs. 3a and 3b) is located just north of the Ma´aloula Canyon and has an outstanding view of the highland Ma´aloula Basin, including the springs

[11] Asouti – Austin 2005.

[12] Conard *et al.* 2006.

Fig. 2: Baaz Rockshelter: a. view from site with Wadi Jaba´deen in front (arrow) (photo: TDASP, 2004); b. site behind fence and Jaba´deen in front of site (arrow) (photo: TDASP, 2004); c. view towards the southeast (photo: K. Deckers, 2005).

and a small lake that existed above the current canyon during the Late Pleistocene.[13] Baaz Rockshelter and Kaus Kozah Cave were investigated by a joint Damascus-Tübingen team, respectively between 1999-2004 and 2004-2006.[14]

At Baaz Rockshelter, seven major archaeological horizons have been identified, each representing periods between roughly 34 kyr BP and the late fifth millennium calibrated BC (figs. 4 and 5). The lowermost strata at Baaz were only reached in the southern area of the excavation and reflect Upper Palaeolithic occupations in the Late Pleistocene. Two radiocarbon dates were retrieved from archaeological horizon (AH) VII and provided ages between circa 34 and 32 kyr BP. In contrast to upper strata of the rockshelter which are mainly anthropogenic, the lower horizons, with the exception of AH V, contained less cultural material and were mainly formed by geological processes. No archaeological features have yet been documented within the lowermost strata (IV, V, VI, and VII), and macro-botanical remains are scarce.

Radiocarbon dates on the charcoal, and artefacts from AH III and II, indicate a more significant phase of occupation in the Late Natufian during the eleventh and perhaps twelfth millennium calibrated BC.[15] It is important to note that this phase of occupation contains an *in situ* floor with a hearth and a built-in mortar in AH IIIb (fig. 6). The lithic artefacts from the Natufian strata suggest that a variety of activities took place, probably ranging from hunting, to preparing meat and plant resources, to the manufacturing of lithic artefacts and wood working.[16] One of the radiocarbon dates of horizon II suggests some intrusions of later occupation remains into this level, although most of the artefacts of horizon II belong to the Late Natufian.[17]

---

[13] Dodonov *et al.* 2007.

[14] see Conard *et al.* 2006.

[15] Conard *et al.* 2006.

[16] Barth 2006; Hillgruber 2010.

[17] Conard 2002.

Fig. 3: Kaus Kozah Cave: a. view looking south with Ma´aloula cliffline in background; b. view towards the site (location indicated with arrow) (photo: TDASP, 2010).

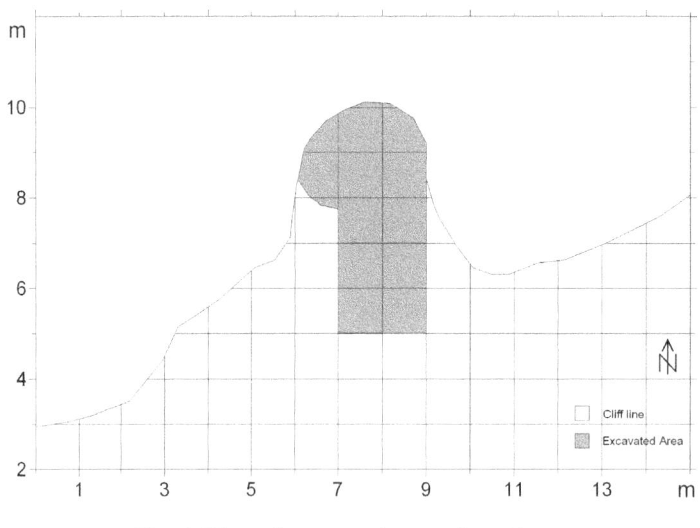

Fig. 4: Plan of excavated areas from Baaz.

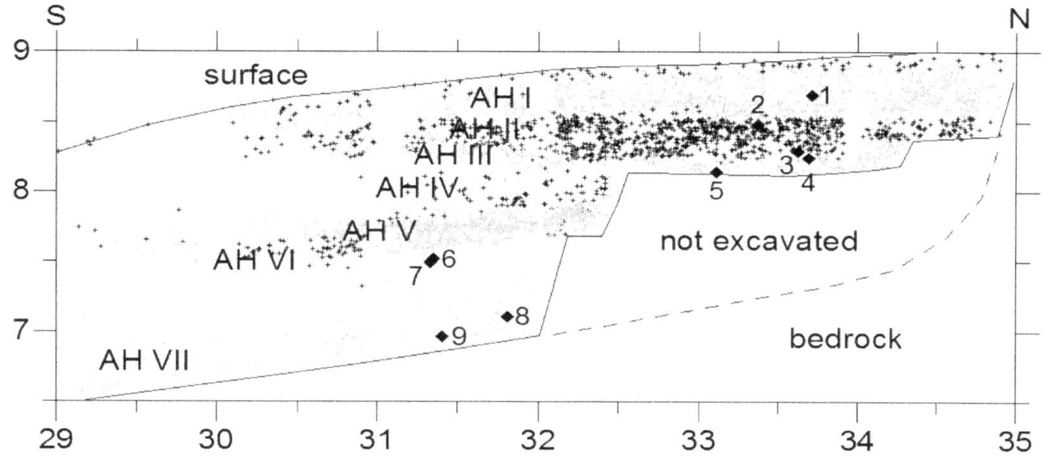

	lab number	m² - find #	AH	material	uncalibrated	calibrated 1 sigma cal BC
1)	KIA11580	20/33 - 163	I	charcoal	5241±35 BP	4216-4203, 4136-4134, 4045-3983
2)	KIA11579	20/33 - 425	II	charcoal	5707±34 BP	4578-4572, 4552-4495, 4469-4464
3)	KIA11578	20/33 - 672	II	charcoal	10667±97BP	10962-10821, 10813-10680, 10500-10492
4)	KIA11577	20/33 - 714	III	charcoal	10942±65 BP	11161-11113, 11074-10935
5)	KIA11576	20/33 - 745	IIIa	charcoal	10470±121 BP	10865-10348, 10284-10178
6)	KIA30307	19/31 -1067	V.1	charcoal	21310+740/-680 BP	n.a.
7)	KIA30308	19/31 - 1068	V.1	charcoal	23040+ 270/-260 BP	n.a.
8)	KIA30309	19/31 - 1403	VII	charcoal	34200+1460/-1240 BP	n.a.
9)	KIA30310	19/31 - 1413	VII	charcoal	32060+600/-560 BP	n.a.

Fig. 5: Profile projection of piece-plotted finds from the excavation units of the 20 East row showing the location and dating results for nine charcoal samples. The five radiocarbon dates younger than 12000 BP were calibrated using CALIB rev4.0, test version6 (Stuiver *et al.* 1998).

Fig. 6: Natufian floor level IIIb with built-in mortar and fireplace (photo: TDASP, 1999).

A radiocarbon sample of the uppermost part AH I provided a late fifth millennium calibrated BC date. However, the lithic artefacts also indicate remains of occupation in horizon I from the earlier PPNA.[18] Although some pockets of charcoal were found within this horizon, no *in situ* structures were found, which could be due to the fact that this horizon is disturbed. Horizon I did not contain any artefacts related to the preparation of food plants, such as mortars or pestles. Most of the lithic artefacts were probably associated

[18] Barth 2006.

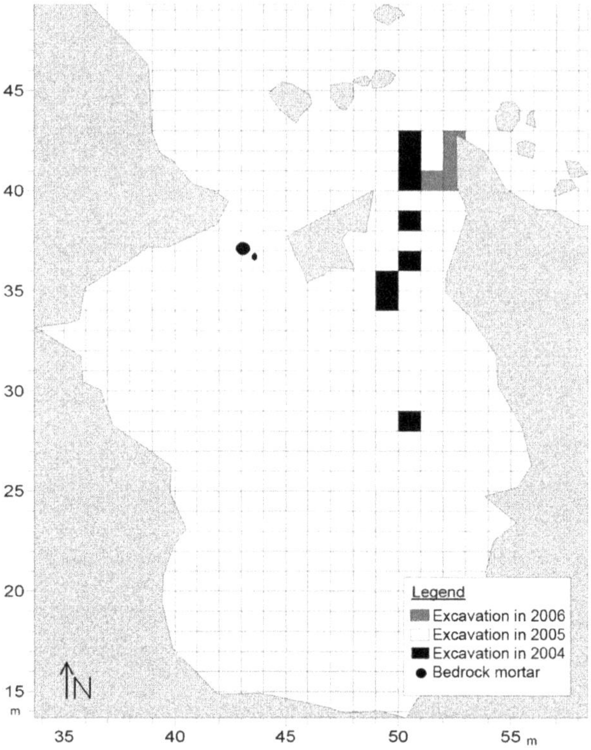

Fig. 7: Plan of Kaus Kozah Cave with excavated areas from 2004 to 2006.

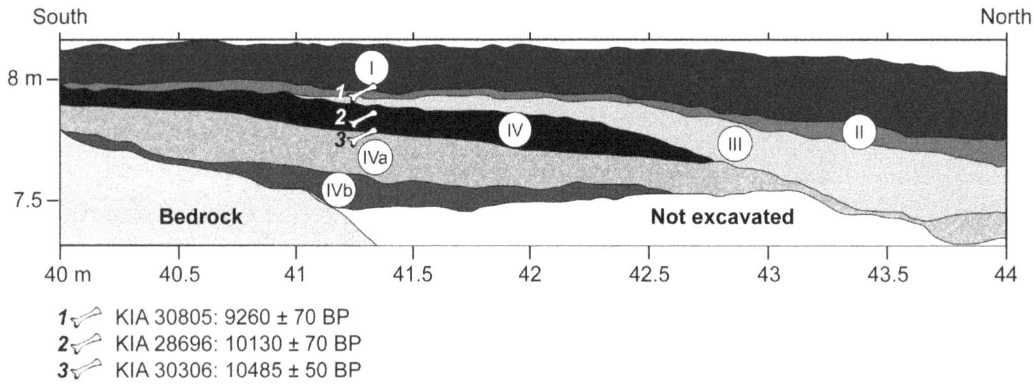

Fig. 8: Kaus Kozah Cave. North-South stratigraphic profile along fifty metre west line showing the location and dating results for three samples. The radiocarbon dates on collagen were calibrated using CALIB rev4.0, test version6 (Stuiver *et al.* 1998).

with hunting, preparing meat, and knapping. The scarcity of tools for preparing plant-foods suggests that the site was not occupied permanently in the Early and Mid Holocene or that plant processing was not a major activity at Baaz.[19]

Kaus Kozah Cave contains four horizons that were defined on the basis of geological changes of the sediment (figs. 7 and 8). The uppermost three horizons seem to contain mainly PPNA artefacts, while the lower horizon (IV) contains Levallois flakes, which are indicative of Middle Palaeolithic occupation. The Levallois flaking technique has often been associated with Neanderthals or early modern humans. Furthermore, within horizon IV human bones belonging to two children were found and

[19] Barth 2006.

dated to the mid eleventh millennium calibrated BC. The excavators suggest that a burial pit must have been dug into the middle Palaeolithic horizon; however, signs for this were not visible during the excavations. Lastly, a date on humic acid from a bone sample from horizon II provided a mid ninth millennium calibrated BC radiocarbon date, but dates on base-soluble humic acid fractions are generally too young. Additionally, several bedrock mortars from Kaus Kozah provide further evidence for a Natufian component at the site.

Generally, the integrity of the horizons at Kaus Kozah is less clear than at Baaz and some degree of mixing took place between horizons I, II, and III. This is particularly evident by the discovery of recent intrusions, such as ceramic, glass, and metal, which are mainly concentrated in horizons I, II, and III.[20] Moreover, during seed and fruit analysis, uncarbonised and a few more recent seed remains (anatomically more recent remains and also a later imported species from the New World) were found in horizons I through III.[21] Additionally, the number of identifiable features at Kaus Kozah is low when compared to the younger layers of Baaz and suggests that disturbances took place at the site.

In summary, while Baaz has many discrete Palaeolithic occupation layers, Kaus Kozah only contains a Middle Palaeolithic occupation layer. Both sites had important occupations at the transition from Pleistocene to Holocene, within the progression from Epipalaeolithic to Neolithic. The artefacts indicate that the main occupation of Kaus Kozah was slightly later than the one at Baaz,[22] with Baaz containing mainly Natufian artefacts and Kaus Kozah having largely PPNA artefacts[23]. Because the main lithic material recovered from Kaus Kozah dates to the PPNA, we will assume most of the charcoals do as well. However, we need to be aware that later intrusions may occur.[24]

The goal of this study is to reconstruct the vegetation in the sites´ surroundings at the transition from Pleistocene to Holocene and to determine what implications this has for the archaeology of the region.

3. Methods

Macrobotanical samples were retrieved via various sampling procedures from almost all investigated strata at Baaz Rockshelter and Kaus Kozah Cave.[25] Flotation samples were taken from Baaz and Kaus Kozah and much botanical material was recovered during dry-sieving and as piece-plotted excavation finds.

Identifications were based on fresh transverse, tangential, and radial fractures at magnifications of 60x, 100x, 200x, and 500x, and were achieved with the use of several wood anatomy atlases.[26]

In total, 11,455 charcoal fragments were identified from thirty-three flotation samples and eighty-four hand-picked samples from Baaz, and 476 fragments from thirty-five samples from Kaus Kozah. So far, only four samples were analysed that were retrieved by flotation (52/42-236, 51/40-311, 20/40-224 and 49/35-120), the others derived from sieving.

4. Results

Tables 1 and 2 contain the results for each of the samples, respectively from Baaz Rockshelter and Kaus Kozah Cave. Table 3 provides an overview of samples from the transition of the Pleistocene to the Holocene at both sites. The hand-picked samples from Baaz were not included within the fragment percentages. They mostly contained *Populus* (poplar)/*Salix* (willow) and were retrieved close to one another, especially in square 19/32 of horizon IIIa. It should be noted that the results of the flotation samples from Pleistocene strata from Baaz derive from a smaller fraction and therefore the results are not completely comparable.

The results from Baaz and Kaus Kozah are similar to one another, with *Amygdalus* (almond) dominating their charcoal assemblages. Because the sites are located in close proximity to each other and in the same environmental zone, this confirms that charcoals from each site are representative of the local vegetation.

---

[20] Conard *et al.* 2006, 200.

[21] Personal communication with S. Riehl, September 2010.

[22] Personal communication with F. Hillgruber, February 2010.

[23] Hillgruber 2010.

[24] Just before this volume was printed, a few new radiocarbon dates were received for Kaus Kozah. They underline a considerable mixing of horizons within the site. For example, a recent *Vitis* seed and a 7th millennium cal BC *Celtis* seed both were included in horizon IV.

[25] No flotation samples could however be retrieved from stratum VII.

[26] e.g. Fahn *et al.* 1986; Gale – Cutler 2000.

Table 1: Identification results of charcoals from Baaz samples.

	Hand-picked samples																		Flotation samples					
	2000, 21/31	2000, 21/31	2000, 21/31	2000, 21/31	2000, 21/31	2000, 21/31	2000 21/31	2000,21/31	2000, 21/31	2000, 19/32	2000, 21/34	2000, 21/34	2000, 21/34	2000, 21/34	2000, 21/34	2000, 21/34	2000, 21/34	2000, 21/34	2004, 19/30	2004, 19/34	2004, 19/34	2004, 19/34	2004, 19/31	2004, 1934
	148	139	109	116	155	153	156	204	152	659	148	113	135	122	108	116	148	99	36	43	77	53	54	37
	IB	I	I	I	IB	IB	IB	I	IB	IA	I	I	I	I	I	I	I	I	I	I	I	I	I	I
Acer				1															1					
Amygdalus			1								7	1	25	7	10	14	15	143	171	343	270	361	193	158
Chenopodiaceae																			15	1	1	5	15	4
Juniperus																			1					
Maloideae																			15	3	4	5	11	29
monocotyledon																						1		
Pistacia non-atl.																			18	11	19	44	28	25
Populus/Salix	13	73			18	1	32	10	101	22									1					2
Rhamnus									2										2	13	2	8	4	9
indet.																			20	8	3	7	20	19

	Hand-picked samples					Flotation samples								
	2000, 21/31	2000, 21/31	2000, 21/31	2000, 21/31	2000, 21/31	2004, 19/34	2000, 19/31	2004, 19/34	2004, 19/34	2004, 19/34	2000, 19/31	2004, 19/31	2004, 19/34	2004, 19/31
	182	220	160	191	205	143	97	188	169	196	120	220	177	177
	II	II	II	II	II	II	II	II	II	II	II	II	II	II
Amygdalus						1506	62	390	966	189	91	71	634	135
Cedrus												2		
Chenopodiaceae						4	43	4	1	1	15	7	4	11
conifer without resin ducts														1
Fraxinus													1	
Juniperus														1
Maloideae									4	5	4	4	8	1
monocotyledon													2	
Pistacia non-atl.						1	6	32	5	5	4	7	14	13
Populus/Salix	1	92	12	18	18				4		1	6	9	6
Rhamnus							1						1	1
Vitis														2
indet.						5	13	23	91	10	9	8	6	1

	Hand-picked samples																											
	2000, 21/31 268	2000, 21/31 301	2000, 21/31 262	2000, 21/31 266	2000, 19/32 584	2000, 19/32 583	2000, 19/32 658	2000, 19/32 635	2000, 19/32 571	2000, 19/32 581	2000, 19/32 568	2000, 19/30 582	2000, 19/32 644	2000, 19/32 590	2000, 19/32 660	2000, 19/32 632	2000, 19/32 578	2000, 19/32 613	2000, 1932 591	2000, 19/32 565	2000, 19/32 648	2000, 19/32 642	2000, 19/32 589	2000, 19/32 626	2000, 19/32 645	2000, 19/32 618	2000, 19/32 631	2000, 19/32 614
	III	III	III	III	IIIA	III	IIIA	IIIA	III	IIIA	III	IIIA	IIIA	IIIA	IIIA	IIIA	IIIA	IIIA	IIIA	III	IIIA	IIIA	IIIA	IIIA	IIIA	IIIA	IIIA	IIIA
Amygdalus										20				3														
Juniperus						14								3									8				2	
Maloideae			9																		1							
Populus/Salix	113	6		28	2	11	38	17	111		17	32	32	20	5	6	27	1	6	2	17	30	3	4	10	2	4	3

	Hand-picked samples (continued)															Flotation samples												
	2000, 19/32 567	2000, 19/32 464	2000, 19/32 539	2000, 19/32 619	2000, 21/32 806	2000, 21/32 836	2000, 21/32 827	2000, 21/31 305	2000, 21/31 313	2000, 21/31 315	2000, 21/31 308	2000, 21/31 309	2000, 21/33 806	2004, 20/32 824	2004, 20/31 434	2004, 20/34 556	2004, 19/34 249	2004, 19/34 230	2004, 19/34 271	2004, 21/34 320	2004, 19/31 259	2004, 19/31 230	2004, 20/34 531	2004, 19/33 771	2004, 19/34 198	2004, 20/31 454	2004, 19/31 304
	III	III	III	IIIA	IIIC	IIIC	IIIC	III	III	III	III	III	IIIC	IIIB	IIIA	IIIA	IIIA	IIIA	IIIA	IIIA	IIIA	III	IIIA	IIIA	III	IIIA	III
Amygdalus				15			1		1		5		10			327	415	633	522	250	193	93	372	54	71	17	84
Chenopodiaceae																46		24	25	4	2	4	9	2	4	2	
conifer with resin ducts																						4					
Fraxinus																			3								
Juniperus																1											
Larix/Picea																						2					
Maloideae																6		13	7	1		1	3	3			3
monocotyledon																		1				1					
Pistacia non-atl.																	3	24	17	3	3	5	2	1	2	1	4
Populus/Salix	1	13	3	1	12	13	23	48		1		1	16	1	4	14	4	85	17		36	6	6	2	2		8
Rhamnus																4		8	7	1		5	1	2	1		
indet.																13	10	37	56		6	23	2	2			

	Hand-picked samples								Flotation samples, small fraction				Hand-picked samples				Flotation samples, small fraction		Hand-picked samples						
	2004, 21/31 351	2004, 20/31 492	2004, 20/31 474	2004, 20/31 524	2004, 20/31 432	2004, 21/32 1094	2004, 21/32 925	2000, 21/32 1026	2004, 19/31 550	2004, 19/31 387	2004, 19/31 354	2004, 19/31 390	2004, 20/31 665	2004, 21/31 419	2004, 19/30 665	2004, 19/30 383	2004, 19/31 656	2004, 19/31 932	2004, 20/34 398	2004, 19/31 749	2004, 19/31 1413	2004, 19/31 1403	15/31 1068	2004, 19/31 1372	2004, 20/31 1133
	IV.1	IV	IV	IV	IV	IV	IV	IV	IV.1	IV	IV	IV	V.1	V.1	V.1	V.1	V.1	V.1	VII	VII	VII	VII	VII	VII	VII
Amygdalus			5		31				20	12	14	25			1			2							
Artemisia monosperma									1	1	2							7							
Chenopodiaceae						4											5		1	1	1	1	1	1	1
Populus/Salix	3	1		1			2	13		2		1	9	1		1		1							
Tamarix										1															
indet.										4															

Table 2: Identification results of charcoals from Kaus Kozah samples. * a new radiocarbon date has shown this is a recent pip.

	50/43	50/43	50/38	49/34	49/34	50/40	50/43	50/43	49/35	49/34	49/35	49/35	50/43	50/40	50/40	50/43	49/35	49/35	49/35	49/35	50/42	50/42	49/35	52/42	52/41	49/35	50/42, 7.693	50/42, 7.639	50/42, 7,669	51/40, 7.632	51/40, 7.844	52/40, 7.881	50/40, near skeleton	50/41, 7.470	20/40, near skeleton
	49	23	133	110	75	31	79	99	118	131	116	115	184	181	89	197	124	122	136	123	98	103	125	236	140	120	135	125	101	311	151	140	227	254	224
	I						II									III											IV								
Acer				1	2					13				1					5				2			5			4						
Amygdalus	41	3	2		10	3		24	1	21	2	6	6	4	3	2	4		2	1		1		83	2	19	1	3		3	15	6	15	1	20
Artemisia monosperma															1														1						
Chenopodiaceae					1					1															1	3			1						
conifer without resin duct				1																															
Fraxinus		27								1																									
Juniperus	1						1																			1					1				
Leguminosae spec.																										1									
Maloideae	5	1	1		1													1			1		1								1				
Pistacia atl.				1																															
Pistacia non-atl.	3		19	2						1												1				2	1		6						
Platanus																										1									
Populus/Salix					1		3	1																	1										
Quercus dec.																										1									
Rhamnus	3		2									1																	1		1				
Rhus																										3									
Tamarix		3																																	
Vitis																	1									3									pip*
indet.		1						4		3	1						3								2	7						1			

	50/42, 7.693	50/42, 7.639	50/42, 7,669	51/40, 7.632	51/40, 7.844	52/40, 7.881	50/40, near skeleton	50/41, 7.470	20/40, near skeleton
	135	125	101	311	151	140	227	254	224
	IV								
Acer			4						
Amygdalus	1	3		3	15	6	15	1	20
Artemisia monosperma			1						
Chenopodiaceae			1						
Juniperus					1				
Maloideae					1				
Pistacia non-atl.	1		6						
Rhamnus			1		1				
Vitis									pip
indet.						1			

Table 3: Summary of fragment percentages of transitional Pleistocene-Holocene samples.

	KK I-III	Baaz III	Baaz II	Baaz I
Acer	8.06			0.05
Amygdalus	66.67	84.26	90.86	80.00
Artemisia monosperma	0.28			
Cedrus			0.04	
Chenopodiaceae	1.67	3.39	2.02	2.19
conifer without resin duct	0.28	0.11	0.02	
Fraxinus	7.78	0.08	0.02	
Juniperus	0.83	0.03	0.02	0.05
Larix/Picea		0.06		
Leguminosae spec.	0.28			
Maloideae	3.06	1.03	0.58	3.58
monocotyledon		0.06	0.04	0.05
Pistacia atl.	0.28			
Pistacia non-atl.	7.78	1.03	1.95	7.75
Platanus	0.28			
Populus/Salix	1.67	5.00	0.58	0.16
Quercus dec.	0.28			
Rhamnus	1.67	0.81	0.07	2.03
Rhus	0.83			
Tamarix	0.83			
Vitis	1.11		0.04	
indet.	5.83	4.14	3.73	4.12

5. Discussion

5.1 Vegetation zones represented at the sites during the Pleistocene-Holocene transition

Besides *Amygdalus*, other typical taxa common to the almond woodland-steppe occur at both sites, such as *Pistacia* (pistachio), Maloideae species, *Rhamnus* (buckthorn), *Acer* (maple), *Juniperus* (juniper), and, possibly, Chenopodiaceae (goosefoot family). Overall, though, it is of particular interest that Kaus Kozah contains a fairly large percentage of *Pistacia*, a fragment of deciduous *Quercus* (oak), and a smaller proportion of Chenopodiaceae. The latter are drought indicators and also occur more frequently in the steppe. This suggests that at the time of Kaus Kozah's PPNA occupation, the environment was somewhat moister than when Baaz had its Natufian occupation. Although the radiocarbon dates from both sites are rather similar (figs. 5 and 8), it is interesting to note that the artefacts indicate that the main occupation phase of Kaus Kozah, represented by PPNA artefacts, was somewhat later than the main occupation phase at the end of the Pleistocene from Baaz, which contains mainly Natufian artefacts. The Natufian occupation from Baaz probably took place during the Younger Dryas, while the PPNA in the Near East typically began afterwards when climatic conditions started to improve. This could explain the somewhat lusher *Amygdalus-Pistacia* woodland-steppe seen at Kaus Kozah.

Besides the woodland-steppe taxa, both sites have a relatively large proportion of hydrophilic vegetation; for example, *Fraxinus* (ash), *Populus/Salix*, and *Vitis* (vine) occur at both sites, while *Tamarix* (tamarisk) and *Platanus* (plane) only occur at Kaus Kozah. This shows that both sites had access to water. This is consistent with the location of both sites near perennial springs. Even today these springs are important sources of water and explain why the towns of Ma'aloula and Jaba´deen are located at these points in the landscape.

Of special interest is the find of *Vitis* at both sites. It suggests that at the end of the Pleistocene/start of the Holocene the distribution of *Vitis vinifera* (grape vine) was somewhat further inland than at present. This may have been due to the different climatic conditions and their associated dislocations of vegetation.

5.2 Possible food plants represented by charcoal

Vitis finds are also of particular interest because they may represent a food plant, such was the case in the Levantine Upper Palaeolithic site of Ohalo II and the Middle Palaeolithic site of Kebara Cave.[27] It is particularly of note to mention that at Baaz (level I), an un carbonised pip was found; The humified pip from horizon IV at Kaus Kozah has been recently dated and appears to be a modern intrusion. The possible role of *Vitis* as nutrition at the site therefore needs to be investigated further.

[27] Weiss *et al.* 2004.

Another plant that may have been used as a food resource at Baaz and Kaus Kozah may have been *Amygdalus*. Unfortunately, no almond shells have yet been found. At many Epipalaeolithic sites wild almonds have been found suggesting their use as food.[28] Wild almonds, although toxic, have a high nutritional value if they are consumed in low quantities and in combination with meat or in combination with simple processing.[29] Several explanations are possible for the lack of wild almonds at the site. The most likely explanation is that the sites were seasonally used; a lack of wild almonds may indicate a settling in spring or early summer. It is also possible, albeit less likely, that the shells were removed elsewhere, or that the entire fruits were consumed in an unripe state.[30]

5.3 Impact of the Younger Dryas on the vegetation

In Deckers *et al.*[31] it was mentioned that the impact of the Younger Dryas was only slightly visible in the anthracological record from Baaz; meaning, that the Natufian and those layers, which also include later artefacts, do not differ greatly from one another. This would have underscored the results from Huleh – that woodland expansion took place but that it did not attain such high proportions as present before.[32] This also fits with the results from Napierala *et al.*, which found salmonid and cyprinid fish bones at Baaz in Natufian layers, indicating a relatively high level of humidity.[33] However, Kaus Kozah, which had its main occupation phase slightly later than what the majority of the artefacts from Baaz layers III to I indicate,[34] has a charcoal assemblage that signifies a somewhat lusher environment compared to Baaz. This is particularly seen in Kaus Kozah's sample of deciduous *Quercus* and in the fact that it contains a larger proportion of *Pistacia* and a slightly smaller proportion of Chenopodiaceae than is present at Baaz III and II. Hence, the charcoal data indicates that during the Younger Dryas both sites were somewhat further away from the zone with oak-Rosaceae park-woodland than immediately afterwards, in the Early PPNA. This has some implications for the availability of wild cereals and subsistence economies of the sites. During the Younger Dryas, the sites seem to have been located just outside the range of the dense stands of wild cereals, while directly afterwards they may have been located somewhat closer. Although the archaeobotanical analysis of Kaus Kozah is still ongoing it is already possible to say that at both Baaz and Kaus Kozah no finds indicate an intensified use of cereals. This may be due to the fact that both sites were hunting posts rather than plant processing sites.

In order to better document the impact of the Younger Dryas on the vegetation, it is important to gain an understanding of the Pleistocene vegetation within the region and how it correlates with the assumed climatic evolution. Regrettably, there are varying reconstructions of the Pleistocene climate in the eastern Mediterranean, some of which reach different conclusions.[35] For example, some consider the Pleistocene in the Near East as extremely dry compared to the Holocene, while others suggest far moister conditions prevailed than at present throughout much of the Pleistocene.

Qualitatively, it is interesting that already in the layer dated to 23 kyr BP and onwards (in layers V and IV), at about the time of the so-called "Last Glacial Maximum," *Amygdalus*[36] was present at Baaz. Yet within the earlier Pleistocene level VII, dated at about 34-32 kyr BP, none were found and only Chenopodiaceae were present. Although there is possible stratigraphic intermixing within level IV at Kaus Kozah, in the form of the previously mentioned mid eleventh millennium calibrated BC burial pit that was not visible during excavations, charcoal samples originating away from this more recent disturbance indicate that in the middle Palaeolithic layers of Kaus Kozah *Amygdalus* was also present.

The charcoals from AH VII at Baaz indicate that climatic conditions were rather dry around 34-32 kyr BP. This has also been suggested by several palaeoclimatic proxies, especially at Soreq Cave.[37] It also fits with the vegetation models of van Zeist – Bottema,[38] which indicate that for several millennia during the last glacial period the interior of the Near East, including the Damascus region, was dominated

[28] Martinoli – Jacomet 2004.

[29] Martinoli – Jacomet 2004.

[30] Hillman 2000.

[31] Deckers *et al.* 2009.

[32] van Zeist – Bottema 1982; Baruch – Bottema 1991.

[33] Napierala *et al.* in press.

[34] Although the radiocarbon dates for Baaz's layers II and I appear to be rather late within the Holocene, most of the artefacts in these layers belong to the Natufian; this might imply that most of the charcoals also belong to the Natufian occupation.

[35] e.g. reviews and discussion in Tzedakis 2007; Enzel *et al.* 2008.

[36] Regrettably, only one fragment exists in the retrieved samples.

[37] see Bar-Matthews *et al.* 1999.

[38] van Zeist – Bottema 1982; 1991.

by *Artemisia* (wormwood), Chenopodiaceae steppe, and desert-steppe. In comparison to AH VII (34-32 ^{14}C kyr BP), AH V and IV charcoals from Baaz suggest that climatic conditions were somewhat moister, probably even from 23 to 21 ^{14}C kyr BP. This early "expansion" of *Amygdalus* (woodland) steppe does not correlate well with the interpretation of the oxygen isotopic data from Soreq Cave that indicates moister conditions only started after approximately 19,000 BP.[39] Perhaps the onset of the Tardiglacial needs to be set somewhat earlier, as the charcoal from Baaz would suggest.

On the other hand, the charcoal data from AH VII to IV at Baaz does not fit well with the alternative reconstruction of the Pleistocene climate claimed by Enzel *et al.*[40] that argues for a pluvial Pleistocene in the Near East.

6. Conclusion

The charcoal from Baaz Rockshelter and Kaus Kozah Cave provide important new data on the development of vegetation from the Pleistocene to the Holocene. The data reveal that rather dry climatic conditions occurred at 34-32 ^{14}C kyr BP but that around 23-21 ^{14}C kyr BP *Amygdalus* steppe had already expanded inland, indicating increased moisture at that time. Although *Amygdalus* steppe occurred during the Younger Dryas in this region, the samples from Kaus Kozah that were associated with PPNA finds suggest that shortly after the Younger Dryas climatic conditions were moister; resulting in a lusher vegetation cover within this region.

Although at first sight the vegetation changes seem to be rather minimal, they could have important implications for where remains of early cultivation sites should be sought. More precisely, while in the Younger Dryas this region seems somewhat further away from the zone of dense stands of wild cereals, shortly after, in the early PPNA, this region was probably located within this zone. The sites of Baaz Rockshelter and Kaus Kozah Cave, nonetheless, do not appear to contain evidence of increased reliance on cereals. This may be due to the fact that both sites mainly served as hunting stations.

Bibliography

ASOUTI, E. – AUSTIN, P.
2005 Reconstructing woodland vegetation and its exploitation by past societies, based on the analysis and interpretation of archaeological wood charcoal macro-remains, Environmental Archaeology 10, 1-18.

BAR-MATHEWS, M. – AYALON, A. – KAUFMAN, A. – WASSERBURG, G.J.
1999 The Eastern Mediterranean paleoclimate as a reflection of regional events: Soreq Cave Israel, Earth and Planetary Science Letters 166, 85-95.

BARTH, M.M.
2006 The lithic artefacts from Baaz Rockshelter, in: N. Conard (ed.), Tübingen-Damascus Excavation and Survey Project, 1999-2005, 25-110.

BARUCH, U. – BOTTEMA, S.
1991 Palynological evidence for climatic changes in the Levant ca. 17000-9000 BP, in: O. Bar-Yosef – F. Valla (eds.), The Natufian culture in the Levant, International Monographs in Prehistory, 11-20.

BAR-YOSEF, O.
1996 The impact of Late Pleistocene – Early Holocene climatic changes on humans in Southwest Asia, in: L.G. Straus – B.V. Eriksen – J.M. Erlandson – D.R. Yesner (eds.), Humans at the end of the Ice Age: The archaeology of the Pleistocene-Holocene transition, 61-76.
2001 The world around Cyprus: From Epi-Palaeolithic foragers to the collapse of the PPNB civilization, in: S. Swiny (ed.), The earliest prehistory of Cyprus: From colonization to exploitation. American Schools of Oriental Research Archaeological Reports, 129-164.
2002 The role of the Younger Dryas in the origin of agriculture in West Asia, in: Y. Yasuda (ed.), The origins of pottery and agriculture, 39-54.

BAR-YOSEF, O. – BELFER-COHEN, A.
1989 The origins of sedentism and farming communities in the Levant, Journal of World Prehistory 3, 447-498.
1991 From sedentary hunter-gatherers to territorial farmers in the Levant, in: S.A. Gregg (ed.), Between bands and states. Center for Archaeological Investigations, Occasional Paper No. 9, 181-202.

---

39 Bar-Matthews *et al.* 1999.

40 Enzel *et al.* 2008.

2002 Facing environmental crisis. Societal and cultural changes at the transition from the Younger Dryas to the Holocene in the Levant, in: R.T.J. Cappers – S. Bottema (eds.), The dawn of farming in the Near East. Studies in early Near Eastern production, subsistence and environment 6, 55-66.

BOTTEMA, S.
1975-1977 A pollen diagram from the syrian Anti Lebanon, Paléorient 3, 259-268.
2002 The use of palynology in tracing early agriculture, in: R.T.J. Cappers – S. Bottema (eds.), The dawn of farming in the Near East. Studies in early Near Eastern production, subsistence and environment 6, 27-38.

CAPPERS, R.T.J. – BOTTEMA, S. – WOLDRING, H. – VAN DER PLICHT, H. – STREURMAN, H.J.
2002 Modelling the emergence of farming: implications of the vegetation development in the Near East during the Pleistocene-Holocene transition, in: R.T.J. Cappers – S. Bottema (eds.), The dawn of farming in the Near East. Studies in early Near Eastern production, subsistence and environment 6, 3-14.

COLLEDGE, S.
2002 Identifying pre-domestication cultivation in the archaeobotanical record using multivariate analysis: presenting the case for quantification, in: R.T.J. Cappers – Bottema – U. Baruch (eds.), The Transition from Foraging to Farming in Southwest Asia, 141-152.

CONARD, N.J.
2002 An overview of the recent excavations at Baaz Rockshelter, Damascus Province, Syria, in: R. Aslam – S. Blum – G. Kastl – F. Schweizer – D. Thumm, D. (eds.), Mauer Schau. Festschrift für Manfred Korfmann, 623-639.

CONARD, N.J. – BRETZKE, K. – HILLGRUBER, K.F. – MASRI, M.
2006 Research in 2005 at Kaus Kozah Cave, in: N. Conard (ed.), Tübingen-Damascus Excavation and Survey Project, 1999-2005, 195-202.

DECKERS, K. – RIEHL, S. – JENKINS, E. – ROSEN, A. – DODONOV, A. – SIMAKOVA, A.N. – CONARD, N.J.
2009 Vegetation development and human occupation in the Damascus region of southwestern Syria from the Late Pleistocene to Holocene, Vegetation History and Archaeobotany 18, 329-340.

DODONOV, A.E. – KANDEL, A.W. – SIMAKOVA, A.N. – MASRI, M. – CONARD, N.J.
2007 Geomorphology, site distribution and Paleolithic settlement dynamics of the Ma´aloula Region, Damascus Province, Syria, Geoarchaeology. An International Journal 22.6, 589-606.

ENZEL, Y. – AMIT, R. – DAYAN, U. – CROUVI, O. – KAHANA, R. – ZIV, B. – SHARON, D.
2008 The climate and physiographic controls of the eastern Mediterranean over the late Pleistocene climates in the southern Levant and its neighboring deserts, Global and Planetary Change 60, 165-192.

FAHN, A. – WERKER, E. – BAAS, P.
1986 Wood anatomy and identification of trees and shrubs from Israel and adjacent regions.

GALE, R. – CUTLER, D.
2000 Plants in archaeology. Identification of vegetative plant materials used in Europe and the southern Mediterranean to c. 1500.

HILLGRUBER, K.F.
2010 The last hunter-gatherers: The Epipalaeolithic in southwestern Syria. Unpublished PhD dissertation, University of Tübingen.

HILLMAN, G.
1996 Late Pleistocene changes in wild plant-foods available to hunter-gatherers of the northern Fertile Crescent: Possible preludes to cereal cultivation, in: D.R. Harris (ed.), The origins and spread of agriculture and pastoralism in Eurasia, 159-203.
2000 Abu Hureyra 1: the Epipalaeolithic, in: A.M.T. Moore – G.C. Hillman – A.J. Legge (eds.), Village on the Euphrates, From foraging to farming at Abu Hureyra, 327-398.

HUSSEIN, K.M.
2006 Climatic characteristics of the Pleistocene and Holocene continental deposits from southwestern Syria based on palynological data, Darwiniana 44.2, 329-340.

KAISER, K. – KEMPF, E.K. – LEROI-GOURHAN, A. – SCHÜTT, H.
1973 Quartärstratigraphische Untersuchungen aus dem Damaskus-Becken und seiner Umgebung, Zeitschrift für Geomorphologie 17.3, 263-353.

MARTINOLI, D. – JACOMET, S.
2004 Identifying endocarp remains and exploring their use at Epipalaeolithic Öküzini in southwest Anatolia, Turkey, Vegetation History and Archaeobotany 13, 45-54.

MEADOWS, J.
2005 The Younger Dryas episode and the radiocarbon chronologies of the Lake Huleh and Ghab Valley pollen diagrams, Israel and Syria, The Holocene 15.4, 631-636.

MUNRO, N.D.
2003 Small game, the Younger Dryas and the transition to agriculture in the southern Levant, Mitteilungen der Gesellschaft für Urgeschichte 12, 47-71.

NAPIERALA, H. – VAN NEER, W. – KANDEL, A.W. – PETERS, J. – UERPMANN, H.-P. – CONARD, N.J.
In press. Fish in the desert? The Younger Dryas and its influence on the paleoenvironment at Baaz Rockshelter, Syria, in: O. Bar-Yosef – F. Valla, F. (ed.), Proceedings of the 2nd Natufian Conference in Paris, 7-11 September 2009.

NESBITT, M.
2002 When and where did domesticated cereals first occur in southwest Asia? In: R.T.J. Cappers – S. Bottema (eds.), The dawn of farming in the Near East. Studies in early Near Eastern production, subsistence and environment 6, 113-132.

STUIVER, M. – REIMER, P.J. – BARD, E. – BECK, J.W. – BURR, G.S. – HUGHEN, K.A., KROMER, B. – MCCORMAC, G. – VAN DER PLICHT, J. – SPURK, M.
1998 INTCAL98 Radiocarbon Age Calibration, 24000-0 cal BP, Radiocarbon 40.3, 1041-1083.

TZEDAKIS, P.C.
2007 Seven ambiguities in the Mediterranean palaeoenvironmental narrative, Quaternary Science Reviews 26, 2042-2066.

UERPMANN, H.-P.
1979 Probleme der Neolithisierung des Mittelmeerraumes. Beihefte zum Tübinger Atlas des Vorderen Orients, Reihe B28.

VAN ZEIST, W. – BOTTEMA, S.
1982 Vegetational history of the Eastern Mediterranean and the Near East during the last 20,000 years, in: J. Bintliff – W. van Zeist (eds.), Palaeoclimates, palaeoenvironments and human communities in the Eastern Mediterranean region in later prehistory. British Archaeological Reports, International Series 133, 277-321.
1991 Late Quaternary vegetation of the Near East. Beihefte zum Tübinger Atlas des Vorderen Orients, Reihe A (Naturwissenschaften 18).

WEISS, E. – WETTERSTROM, W. – NADEL, D. – BAR-YOSEF, O.
2004 The broad spectrum revisited: Evidence from plant remains, Proceedings of the National Academy of Sciences of the United States of America 101(26), 9551-9555.

2. Spatial analysis of Early Bronze Age settlements and resource exploitation in the Upper Khabur Basin (NE Syria)

Katleen Deckers and Philipp Drechsler

1. Introduction

Satellite analysis is a powerful tool to detect archaeological sites and landscapes in the Near East. Geographic Information System (GIS) applications have allowed for an increased understanding of settlement and resource exploitation patterns. Several studies on this subject have in recent times been published for the Upper Khabur, the focus area of this study. They range from seeking to detect hollow ways as a means of understanding contact routes,[1] to using rank size analysis for understanding the hierarchical structure of the settlements;[2] or even deal with calculating the necessary land size to provide sustainable agricultural exploitation.[3] Recently, GIS-based settlement studies in this area have incorporated agent-based models to explore the behaviour of small communities and their associated economic systems.[4]

The present study investigates settlement patterns and land exploitation at the Upper Khabur Basin in northeastern Syria with a chronological focus on the Early Bronze Age (EBA). A variety of methods will be applied in order to gain better understanding of the socio-political and environmental factors that may have played a role in the settlement history and distribution.

2. The database

A study area was chosen within the Upper Khabur based on the availability of satellite images. It covers an area between Wadi Avedji in the west and Wadi Rumelan in the east. Within this study CORONA images with a resolution of roughly two to three metres were analysed to identify archaeological sites.[5] Many sites in this area appear as settlement hills, also called *tells*. A total of 908 tells were mapped within the study area.[6]

The database of sites contained fifty-four sites located close to one another. It was decided that sites of the same period within a distance of 350 metres could not have functioned as independent sites; therefore, they were mapped as one site for most of the analysis (fig. 1), but not for our land exploitation calculations.

Although a large amount of survey work was undertaken in the Upper Khabur area over the past sixty years,[7] not all tells were systematically surveyed. Tells in the western Upper Khabur region were typical multi-period occupations that started at least in the Ubaid or early Late Chalcolithic Period and attained their maximum settlement area during the third millennium BC. However, they show relatively little occupation on the main mounds after the late third millennium BC.[8] Nevertheless, in the Tell Leilan region (eastern Upper Khabur), recent surveys indicated that the main settlement peak was not in the mid third millennium BC but in the early second millennium BC (early Middle Bronze Age).[9] However, for several of our analyses we will assume that all mapped tells had their maximal expansion in the third millennium BC. This results in a slight overestimation of the number of sites for the Leilan survey region for this period,

---

1 Ur 2003.

2 Wilkinson 1994; Goldhausen – Ricci 2005; Risvet 2005; 2008.

3 Wilkinson 2003, fig. 6.18; Deckers – Riehl 2008.

4 Wilkinson *et al.* 2007.

5 "The negatives from the following missions were processed and printed on photo paper: D025-055 1105-1FWD, 5 Nov 68, D025-011, 1102-1 AFT, 11 Dec 67 and D025-003, 1108-1AFT, 6 Dec 69. They were subsequently scanned and geo-referenced. Landsat-7ETM+ three band maps at 14.3 m resolution were used for this.

6 It should be noted that since the Deckers – Riehl 2008 publication, Tell Brak has been mapped differently because in the original version it proved difficult to draw the boundaries of the site accurately solely based on visual interpretation. Due to the fact that archaeological surveys have since been executed in this area, it was decided to adapt the site contours to the illustration presented in Wilkinson *et al.* 2001. Small tells surrounding Brak that were not visible on the Corona images were, however, not mapped.

7 see e.g. Van Liere – Lauffray 1954; Meijer 1986; Lyonnet 1996; 1997; 1998; 2000; Wilkinson 2000; 2002; Wilkinson – Barbanes 2000.

8 Wilkinson – Tucker 1995, 51; Wilkinson 2002.

9 Risvet 2005.

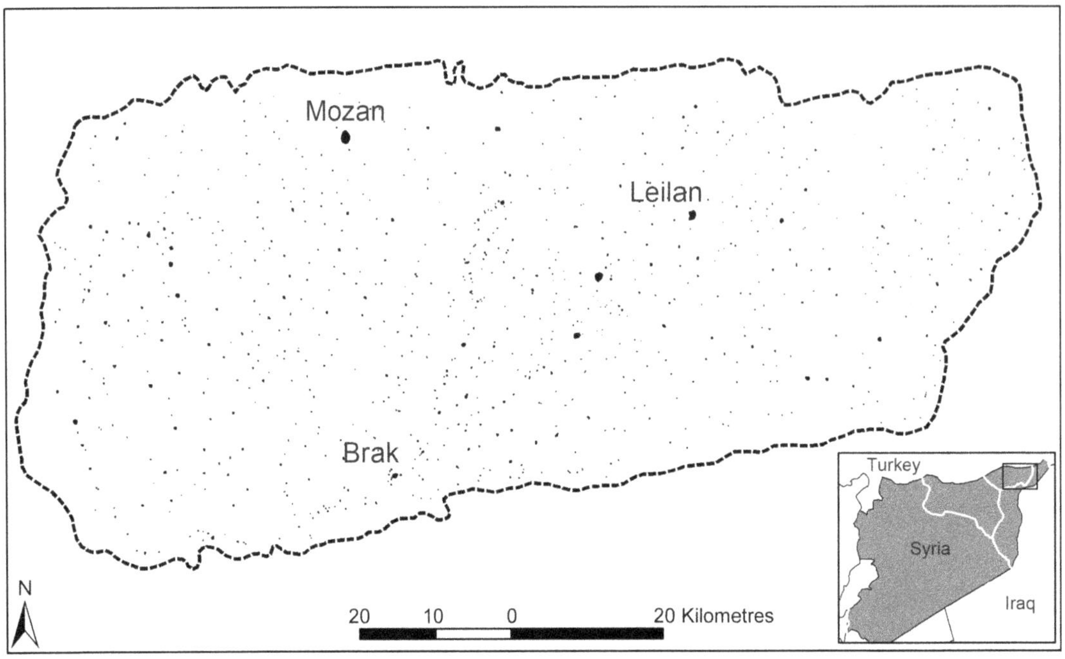

Fig. 1: Map of Upper Khabur Basin with tells and lower towns depicted based on CORONA satellite images.

but will likely not have a great impact on the overall settlement picture. In the section that follows a more detailed insight is gained into the occupation history of the centres within the research area.

3. Central places and settlement hierarchy in the Upper Khabur

Based on site size, the central places of the Upper Khabur can be determined as depicted in Fig. 2. In this map, the site sizes were divided in sites larger than 100 hectares, sites between thirty and 100 hectares, sites between thirteen and thirty hectares, and sites less than thirteen hectares. All sites larger than thirteen hectares can be considered as centres. There are seventeen centres and 837 villages. Some of them have been excavated, while others have only been surveyed, and it is therefore important to have a closer look at their occupation phases. This will also demonstrate that the settlement history is not as static as we assume and that mapping sites by remote sensing is sometimes problematic.

3.1 Urban centres larger than 100 hectares

Mozan, characterised by an upper and lower town, is the only site within the study area that covers more than 100 hectares (fig. 2). Its beginnings are as yet unknown, but the site originated during at least the early part of the fourth millennium BC.[10] Additionally, a major occupation phase dates back to the second half of the third millennium BC. Unlike many other sites in this area, Mozan was continually occupied at the end of the third millennium BC and has also been identified as the important late third millennium Hurrian capital of Urkish. By around 2300 BC it was probably an ally or vassal of the Akkadian Empire. Continued occupation at the site between 2200 and 1900 BC may relate to the newly established kingdom of Urkiš and Nawar. During the second millennium BC the city was passed on to the rulers of Mari and became a vassal; it was ultimately abandoned about halfway through this millennium. Archaeological survey indicates that the main occupation phase of the lower town was in the Early Jazira III Period (ca. 2700-2250 BC).[11]

3.2 Urban centres between thirty and 100 hectares

A total of three different sites were found to each cover areas of between thirty and 100 hectares: Tell Leilan, Farfara, and Khirbat Diba (fig. 2).

The earliest evidence for occupation at Leilan (fig. 2) dates back to approximately 5000 BC and is mainly concentrated on what is referred to as the *citadel*, or the main mound (measuring fifteen hectares). By about 2600 BC, Leilan grew into a large city of roughly ninety hectares. At this time a lower town, with

[10] Buccellati – Kelly-Buccellati 1988.

[11] Pfälzner – Wissing 2004.

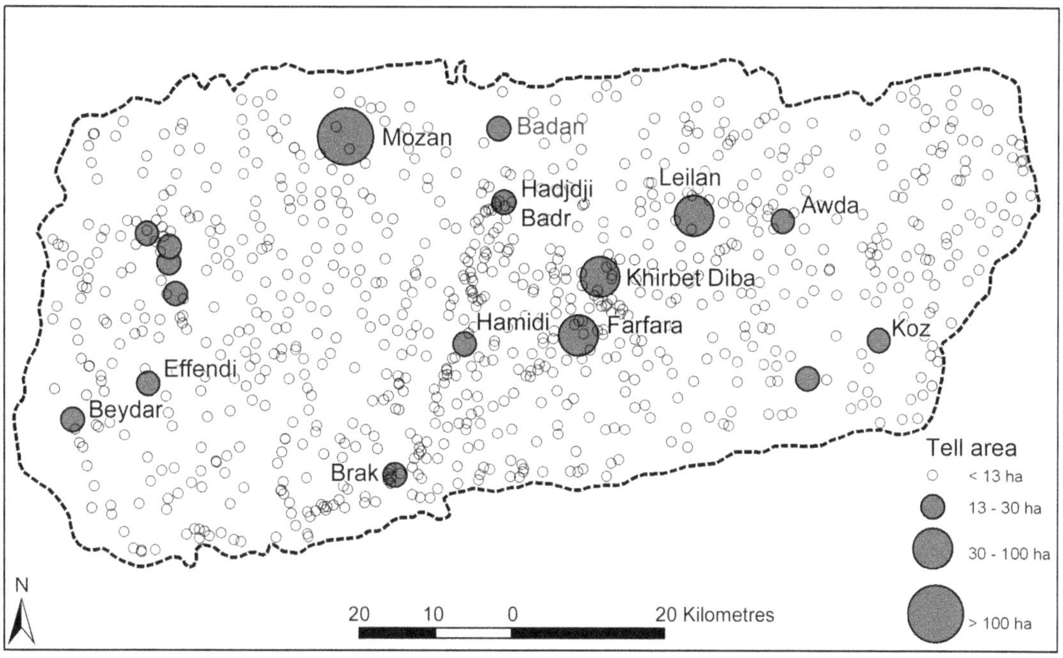

Fig. 2: Central places of the Upper Khabur Basin based on site size as mapped from Corona satellite images.

an associated city wall and ditch, was established.[12] Later, at around 2300 BC, the site came under Akkadian control and some reorganisation took place.[13] Shortly after, by about 2200 BC, Leilan was abandoned for a few hundred years like many other sites in the area.[14] After the conquest of the region by the Assyrian king Shamshi-Adad I (1813–1781 BC), the site of Tell Leilan (renamed Shubat-Enlil) was revived again and it became capital of his northern Mesopotamian kingdom. In the city a royal palace was built and a temple acropolis to which a straight paved street led from the city gate. There also was a planned residential area and renewed construction works took place on the city wall.[15] The city prospered until King Samsu-Iluna of Babylon sacked it in 1726 BC. Shubat-Enlil never regained its power afterwards. It subsequently became a provincial Hurrian town in the kingdom of Mitanni.[16]

Tell Farfara is located eighteen kilometres southwest of Leilan (fig. 2). It was located approximately midway between Leilan and Brak. Its initial growth occurred after the rise of these two cities. In the phase between 2300 and 2200 BC, while other centres were shrinking in size or entirely abandoned, Farfara suddenly grew to nearly forty hectares. These population changes may be attributed to the Akkadian presence in northern Mesopotamia. Between 2200 and 1900 BC, Farfara was abandoned just like many other centres in its surroundings. From 1900 through 1700 BC, Tell Farfara expanded to roughly ninety hectares.[17] For this reason, Farfara became a likely candidate for a Mitanni capital from 1500-1300 BC, either representing Ta'idu or Waššukanni.[18] [19]

The site seen northeast of Tell Farfara can most likely be identified as Khirbat Diba, so named and surveyed by Meijer[20] (fig. 2). Surveys indicated it is probably not a Bronze Age site but rather representative of Byzantine and Islamic periods. Hence, the inclusion of this site into the database indicates that we have an overestimate of sites for the third millennium BC.

[12] Risvet 2005.

[13] Weiss *et al.* 1993.

[14] Weiss *et al.* 1993; Staubwasser – Weiss 2006.

[15] Risvet 2005.

[16] Weiss *et al.* 1993.

[17] Risvet 2005, 121.

[18] Risvet 2005, 160.

[19] Taidu is one of the Mittanni capitals. The majority of the Middle Assyrian texts associate Taidu with toponyms in the Habur. The Dur-Katlimmu itinerary, our best evidence for historical geography, places it a day's travel from Wadi Jaghjagh and two days from Wadi Magrisi (Röllig 1984). Wäfler *et al.* (1985) and Wäfler (1995; 2001) locate it at Hamidi, but Guichard (1994, 244) suggests—based on evidence from Mari—that Farfara is a more likely candidate.

[20] Meijer 1986.

The Leilan survey results indicate that we also omitted one centre, Mohamed Diyab, because it was not visible on our satellite images. This site was founded in the phase between 3000 and 2800 BC as a small village of 3.7 hectares.[21] Between 2800-2650 BC it grew to about seventeen hectares.[22] Its status as a small town changed between 2500-2300 BC when it blossomed in to a small city of fifty hectares.[23] Subsequently, it decreased in size again between 2300 and 2200 BC, to a mere ten hectares. A further reduction in Mohamed Diyab's size occurred in 2200 to 1900 BC,[24] while between 1900 and 1700 BC, it expanded again to at least thirty-five hectares. Ultimately, Mohamed Diyab became largely abandoned by 1700-1500 BC.[25]

3.3 Urban centres between thirteen and thirty hectares

One of the sites site that was mapped within the thirteen to thirty hectare size class is Tell Brak (fig. 2). The site was occupied between the sixth and second millennium BC and had subsequently less massive settlements in the Hellenistic, Roman, and Islamic Period.[26] Size determination using satellite images is however problematic at Brak due to the many small mounds surrounding the site. It is unclear whether these mounds belong to the major tell or whether there are various occupation gaps between the central tell and these smaller sites (fig. 2). In earlier work,[27] we identified Tell Brak as the major site in the area covering 230 hectares. However, Wilkinson who also undertook surveys in the Brak area mapped the major tell as much smaller in size but with many associated smaller surrounding tells.[28] For this reason, we decided to remap the site according to Wilkinson's observations. As a result, Brak now has been classified as a site between thirteen and thirty hectares. Nonetheless, recent survey data indicates that this "minimized" mapped status especially applies to the 4100-3900 BC phase. From 3900-3400 BC and onwards the spatial separation between different settlement clusters at Brak diminished.[29] At around 3000 BC Brak significantly reduced in size similar to many other sites in the region. By 2500 BC Brak once again grew into an important centre.[30] Survey, however, indicated that Brak now showed a different urban structure than previously seen. The outer mounds were abandoned and Brak appears to have had a lower town to the south of the central mound. Based on survey data the site has been estimated to have covered seventy hectares at this time.[31] Texts from 2400 BC indicate Brak (then referred to as Nagar) as one of the three most important states of Upper Mesopotamia alongside Ebla and Mari.[32] There is evidence for an Akkadian intervention circa 2300 BC and the 23rd century BC Naram Sin Palace at the site indicates tribute being paid to the Akkadians at that time.[33] At the end of the EBA the site shrinks in size. By the early second millennium BC the site remained relatively small with at least two areas of houses restricted to the highest ridge of the mound. Finally, in the latter half of the second millennium, the still relatively small area occupied contained a monumental Late Bronze Age (LBA) palace, a Mitanni temple (ca. 1500-1360 BC), and a continuation of the sequence of domestic occupation.[34]

There are several other centres within this size class. One of them is the excavated site of Tell Beydar (fig. 2). This town was founded around 2900 BC and consists of one tell with an accompanying lower town.[35] Beydar was flourishing and probably independent by about 2500 BC, then it fell under the control of Nagar at circa 2400 BC.[36] During the reign of Sargon (ca. 2350 BC), Tell Beydar was controlled by the Akkadians.[37] This period simultaneously marked the beginning of its rapid fall, where its size diminished

[21] Lyonnet 1990; Risvet 2005, 57.

[22] Risvet 2005, 58.

[23] Lyonnet 1990; Risvet 2005, 60.

[24] Risvet 2005, 120.

[25] Risvet 2005, 123.

[26] Matthews 2003.

[27] e.g. Deckers – Riehl 2008.

[28] 2003, 112.

[29] Ur *et al.* 2007.

[30] Matthews 2003; Ur in press.

[31] Ur in press; Ur *et al.* in press.

[32] Steele *et al.* 2003.

[33] Sallaberger 2007, 425.

[34] Ur *et al.* in press.

[35] Milano *et al.* 2005.

[36] Lebeau – Suleiman 2005.

[37] Van der Stede 2005.

considerably, becoming less than one hectare. Sometime around 2100 BC Beydar was abandoned. A short re-occupation took place in the 2nd and 1st century BC.[38]

Another well-known urban centre with an associated lower town is Tell Hamidi (fig. 2). Only the tell site was mapped here because the lower town is not clearly visible on the satellite images. The tell was already occupied during the third millennium BC. Nevertheless, in the Mitanni Period, the site developed into a major centre when it likely became a capital.[39] The town was rebuilt in the middle of the second millennium BC and it was also then that the lower town was constructed, covering an area of 245 hectares.[40] Due to the fact that our investigations only focus on the third millennium BC it is unproblematic that the lower town was not mapped, as it does not date into this millennium.

Tell Effendi is a surveyed urban centre eleven kilometres northeast of Beydar (fig. 2). It consists of a tell that was occupied during the EBA. It also had a lower town settlement in the LBA and Iron Age, however, it was not mapped here.[41] Additionally, the site area had minor occupations from the Roman, Parthian, Byzantine, and Islamic Periods.[42] While Tell Effendi was mapped by Ur – Wilkinson[43] at between seven and ten hectares, we measured it as thirteen hectares.

Another mapped tell can probably be identified as Awda (Oda); it covers an area of twenty-one hectares (fig. 2). According to surveys,[44] the site was occupied between 2800 and 2200 BC, deserted in the phase of 2200-1900 BC and again re-inhabited between 1900 and 1700 BC. It is of note that Risvet mapped it much smaller than we did. She cited the site as – at its largest – five to ten hectares in the 2500-2200 BC phase.[45] Nevertheless, from her work it is not known as to whether she also included the possible lower town in the survey.

Tell Badan is another site that fits this size class, but according to survey data it would not have had its major occupation phase within the third millennium BC (fig. 2). However, it had a major occupation phase in the Halaf Period and minor occupation phases in the Chalcolithic, EBA, and Middle Bronze Age. By the LBA, Iron Age, and onwards the site had substantial occupations once more.[46]

Finally, one site south of Badan and located on the Wadi Jaghjagh, was mapped at a size of fifteen hectares. It can most likely be identified with site 171, Tell Hadjdji Badr, surveyed by Meijer[47] (fig. 2). Here, Meijer found evidence for EBA and MBA occupations.

In addition, along a tributary of the Wadi Khanzir, four un-investigated ancient centres have been found located very close to one another (fig. 2). Neither their names, nor periods of occupation are known. Furthermore, in the southeast of the research area two centres were found that represent un-investigated tells (fig. 2); one of them can be identified as Tell Koz, while the other more southwestern site remains unidentified.

4. Site hierarchies based on site size

Rank size analysis is a well-established method to investigate settlement systems. Developed by geographers to characterise the population distribution across the settlements in a specific region,[48] several requirements have to be fulfilled for its successful application. Not only must the exact size of settlements be known, but also the borders of the settlement system under study. As both preconditions are hardly fulfilled in our dataset it was decided to omit rank size analysis. More specifically, in our research the exact territory boundaries for the third millennium BC are unknown, and the rank-size graphs fluctuated greatly because the sizes of the larger centres were sometimes difficult to map. Nevertheless, conclusions about site hierarchies in the Upper Khabur can be drawn from the spatial distribution of sites.

In Fig. 2, settlements that are cited as centres are distributed somewhat unevenly across the plain. In the Tell Leilan region, for example, Farfara and Leilan – two large centres – are located close to one another;

---

[38] Galán 2005.

[39] Wäfler *et al.* (1985) and Wäfler (1995; 2001) locate Taidu at Hamidi, but Guichard (1994, 244) suggests—based on evidence from Mari—that Farfara is a more likely candidate.

[40] Wäfler 1990, 2001.

[41] Ur – Wilkinson 2008; Wilkinson 2000.

[42] Lyonnet 2000, 35.

[43] Ur – Wilkinson 2008.

[44] Risvet 2005.

[45] Risvet 2005.

[46] Lyonnet 2000, 37 – site 70.

[47] Meijer 1986.

[48] Drennan – Peterson 2004, 533.

while between Mozan in the north and Brak in the south, only small villages occur. For the northern Jezira, a three-tiered settlement hierarchy has often been indicated based on site size. One example is the Tell al-Hawa region in northern Iraq just east of our study area. This locale has one major centre up to sixty-six hectares, second-rank sites between ten and twenty hectares, third-rank sites less than ten hectares in size, as well as secondary centres that have a distance of less than ten kilometres from the main centre.[49]

The situation in the Mozan region appears to deviate somewhat from the aforementioned picture. Mozan stands as the only extremely large centre but lacks secondary centres from the thirteen to thirty hectare size class within a radius of twenty kilometres. The nearest centre within the size class of a secondary centre could be Tell Badan, roughly twenty kilometres east of Mozan. However, as mentioned above, survey indicated that this site was not a major EBA centre. Thirty kilometres to the southwest are more sites that classify as secondary centres, which are not located near a primary centre. Hence, if Tell Badan and the other secondary-sized centres do not belong to the settlement system of Mozan, it looks as if there was only a two-tiered settlement system there. The lack of larger secondary centres in the Mozan surroundings may explain why this centre may have become so extremely large; more specifically, it grew in size because the villages contributed surplus production to it while, in contrast, larger secondary centres generally consume most of their production.[50]

However, as mentioned before, ancient texts indicate that Beydar, for example, needs to be placed within the settlement system of Brak in the 24th century BC.[51] This implies that the settlement system zone of the major centres in this area is somewhat larger in the Upper Khabur than in, for example, the Tell al-Hawa region at its east. Nonetheless, it is also likely that Beydar was a secondary centre of Brak only during the Akkadian phase.

In order to gain insight into the possible minimal sizes of settlement systems, it is important to investigate what amount of land needed to be under cultivation to feed the people of these settlements.

5. Resource exploitation

Based on mapped tell and lower town areas, Deckers – Riehl[52] calculated the sustainable areas for the eastern Upper Khabur Basin region during the EBA using realistic population density values, food consumption patterns, and agricultural yields within this climatic zone.

In this study we assumed that all tells and lower towns reached their maximum expansion during the EBA. It was further postulated that tell sites supported a significant sedentary population. An extrapolation of food consumption patterns from 1970s Syria to the archaeobotanical data made it possible to calculate land needed to feed the people living in this area. This was possible through the use of yield data from FAO for the 1961-1970 period and by taking settlement size into account. For the required amount of seed, an extra 6% was added.[53]

Assuming a population density of 200 persons/hectare within the settlements,[54] and depending on the archaeobotanical data used for the calculations, the area under permanent cultivation was estimated to cover around 20% of the region under investigation (fig. 3b). Furthermore, if one assumes population densities of 400 persons/hectare, roughly 40% of the area would have been under cultivation (fig. 3d). Hence, it is clear that the area as a whole could have been self-sufficient and that there was still even room for pastures and nature, unlike what is often suggested.[55]

Moreover, sustaining areas that show substantial overlapping may indicate dependency, while those with little overlap suggest economic self-sufficiency.[56] The radius of the sustaining area of Mozan is about 7.3 kilometres (assuming population densities of 300 persons/hectare) (fig. 3c). Within this radius sixteen small villages occurred, suggesting that they probably were dependent on Mozan or, more precisely, contributed surplus production to this centre. Likewise, Leilan and Farfara, with respectively five and 3.6 kilometre radii of sustaining area, both had several villages within their agricultural sustaining area.

It is of note that along the whole course of the Wadi Jaghjagh considerable overlap of sustaining areas exists. This suggests that if all settlements were occupied at the same time, not enough land was available within a manageable distance from the village given that people depended on agriculture alone. The Jaghjagh being the largest of all streams was perhaps navigable and therefore could have been an important

[49] Wilkinson 1994.
[50] Wilkinson 1994, 501.
[51] Sallaberger 2005.
[52] Deckers – Riehl 2008.
[53] detailed in Deckers – Riehl 2008.
[54] see Kramer 1980.
[55] Wilkinson 2003, 6.18.
[56] Stein – Wattenmaker 1990.

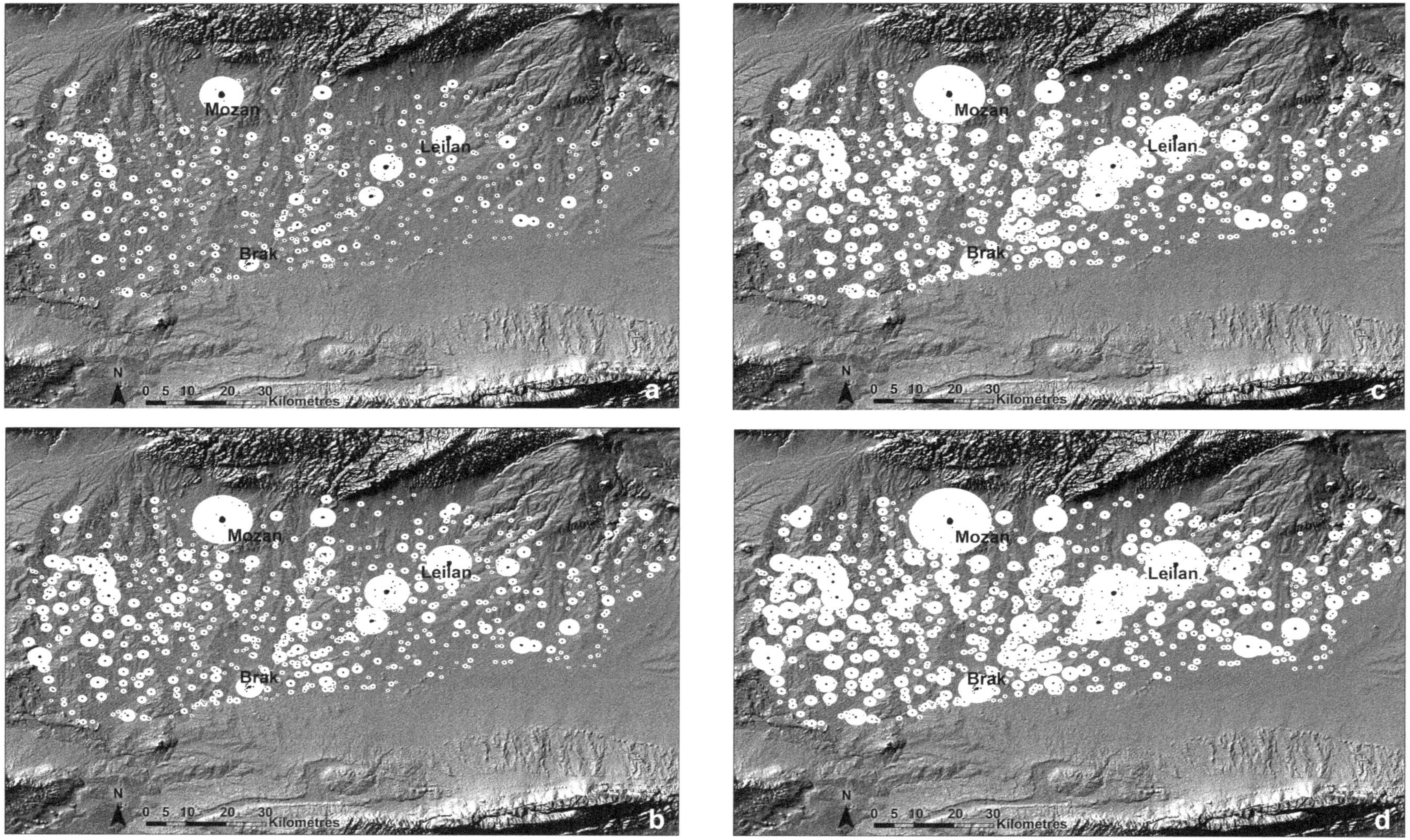

Figs. 3a-d: Estimation of agricultural land necessary to support inhabitants according to proportions established from archaeobotanical data at EBA Tell Mozan and based on site density assumptions for: 100 persons/hectare (fig. 3a), 200 pers/ha (fig. 3b), 300 pers/ha (fig. 3c), and 400 pers/ha (fig. 3d).

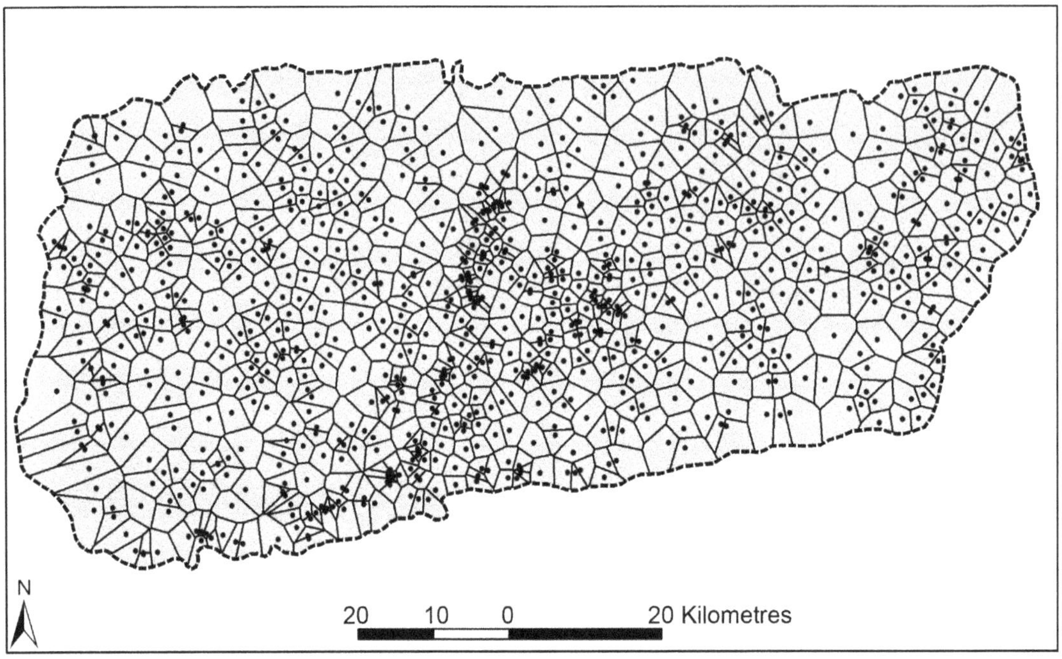

Fig. 4: Thiessen polygons created for the area based on the distance from sites.

link for trade. However, along other streams such as, for example, the Wadi Avedji or Wadi Khanzir, the sustenance zone of one village ended where the next one began. This spacing of villages suggests self-sufficiency for these settlements and may indicate that no surplus was delivered to a major centre.

6. A comparison of calculated resource exploitation zones with territories drawn as Thiessen polygons

Thiessen polygons applied to settlement archaeology are often used to identify the size of the zone of influence for central places.[57] The area contained in each polygon is closer to the point on which the polygon is based than to any other point in the dataset. The assumption is that places closer to a site will likely enjoy a greater amount of interaction than sites further afield.

Table 1: Results of statistical analysis (ANOVA) recognizing correlations between specified factors and site distributions.

Correlation	F-value	Interpretation
settlement area and rainfall	0.6898	no
settlement area and TWI	3.3825	no
normalised settlement area and rainfall	0.9944	no
normalised settlement area and rainfall	0.8998	no
surface of Thiessen polygons and rainfall	2.7339	no
surface of Thiessen polygons and TWI	0.0185	no
sum of town population and rainfall	0.0983	no
sum of town population and TWI	0.2291	no
proportion of town population and rainfall	0.3285	no
proportion of town population and TWI	0.174	no
sum of village population and rainfall	5.8342	yes, more people live in villages in areas with less rainfall
sum of village population and TWI	11.1616	yes, more people live in villages where high TWIs occur
proportion of village population and rainfall	0.0984	no
proportion of village population and TWI	0.3036	no
number of tells and rainfall	1.9252	no
number of tells and TWI	10.5927	yes, more tells in wetter areas
normalised number of tells and rainfall	4.3962	no
normalised number of tells and TWI	8.7377	yes, more tells in wetter areas

[57] Hodder – Orton 1976; Danks 1977; Angell – Moore 1984.

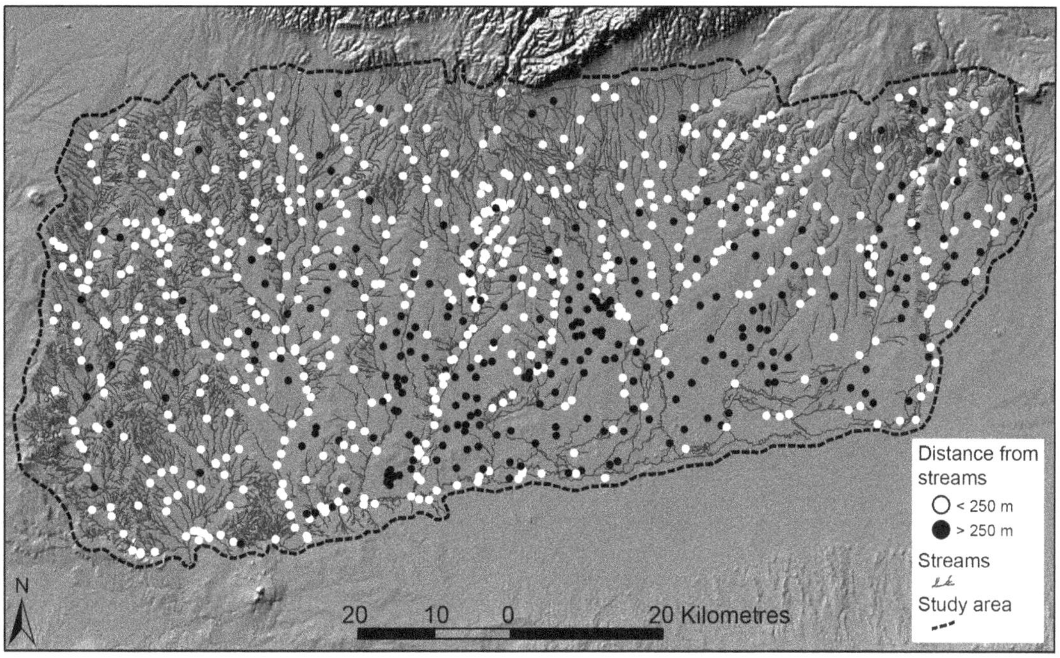

Fig. 5: Map of the Upper Khabur area with analysis of tell locations in relation to streams.

Thiessen polygons were generated for the study area and potentially could provide a general approximation of the extent, shape, and orientation of the spheres of influence or territory around recorded settlements. As can be seen in Fig. 4, there is a significant edge effect along the border of the study area that does not represent the border of the settlement system. Based on this irregularity, we discarded all Thiessen polygons that intersect the boundary of the study area. Unfortunately, no statistically significant correlation was found between the size of the Thiessen polygons and the size of the corresponding site (tab. 1). This indicates that the Thiessen polygon model does not function well in describing the size of territories in this area. Furthermore, the Thiessen polygon model is never capable of generating what was already shown – some secondary sites are located within the resource exploitation zone of primary sites. The distribution of territories and spacing of sites in the area is therefore governed by other factors than distance from a village. For this reason, we need to search for other environmental and socio-political factors that cause this observed distribution.

7. Exploring moisture availability as a steering factor for the settlement patterns

For this study, tells were considered to be located along a stream if they were found less than 250 metres from a current or relict streambed. Of the 854 tells identified as independent settlements within the research area, 68.7% (587) were positively identified as located along an identifiable (relict) stream, while 31.3% (267) were not (fig. 5). It should be mentioned that there may well be some (relict) streams that were not mapped and, therefore, that even more tells may have been located along (relict) streams, as is also suggested by unmapped depressions on the SRTM images. In light of this, our calculations represent a minimum. Moreover, the Wadi Radd area is generally swampy and lacks clearly identifiable channels so that the moist context of some tells may be slightly underrepresented.[58] Water courses constituted a major axis along which human settlement in the Upper Khabur was founded (fig. 5).

Another way to gain insights in to the role of moisture on the settlement pattern is by calculating a topographic wetness index using Beven and Kirkby's formula[59]:

$$TWI = \ln(A_s/\tan \beta)$$

[58] It, however, needs to be investigated as to whether waterlogged conditions also occurred in this area during the Early Bronze Age.

[59] Beven – Kirkby 1979.

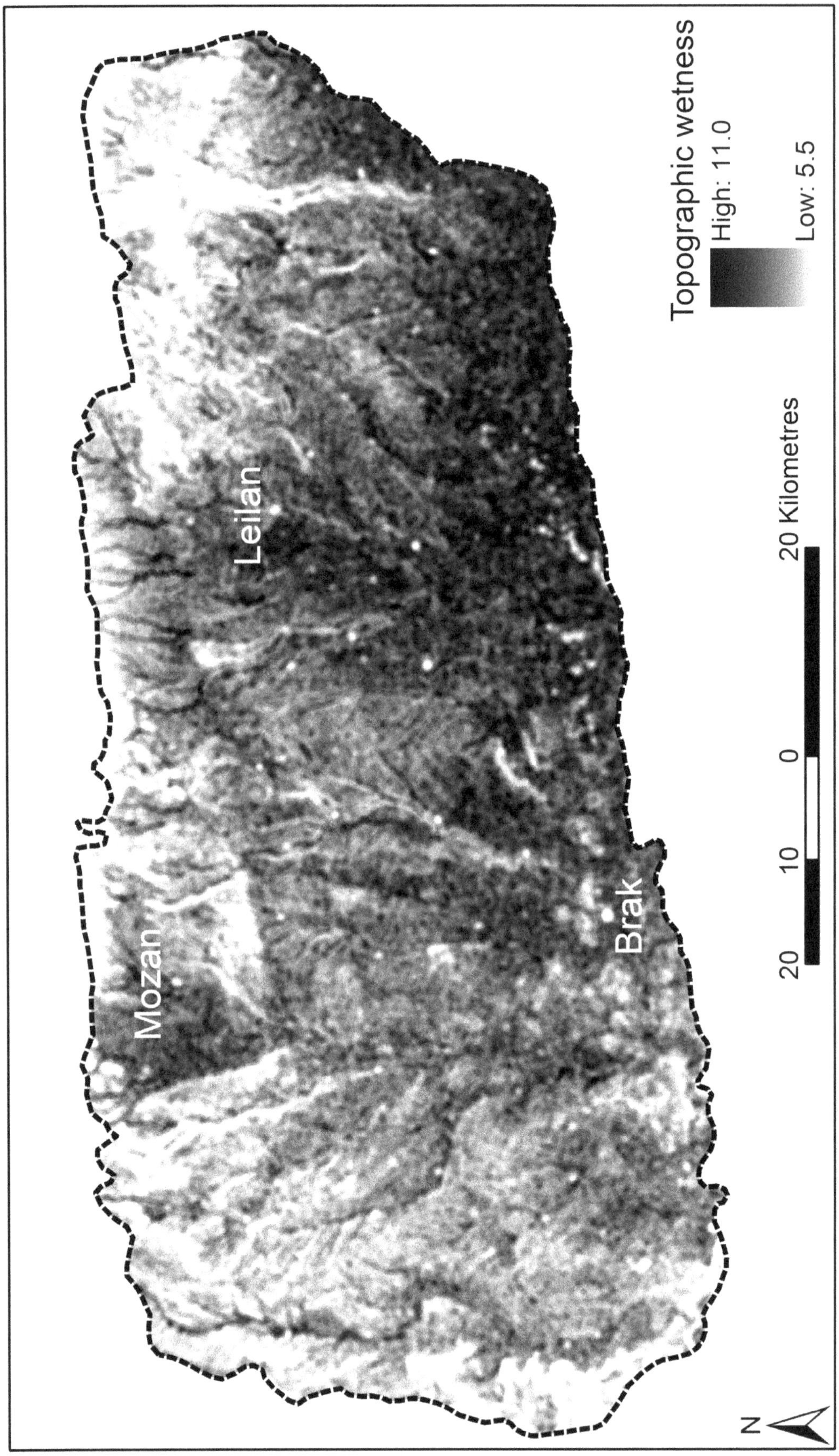

Fig. 6: TWI index map for the Upper Khabur area.

Where A_s is the upslope area and β the slope of a given grid cell in a digital elevation model. The TWI predicts zones of water saturation on the basis of upslope area and slope;[60] hence, high TWI values indicate the concentration of water and thus the spatial pattern of soil moisture availability within a landscape (fig. 6).[61] We could observe a low but significant correlation between the number of tells and the topographic wetness index (tab. 1), with more sites in moister areas. However, there is no correlation between the occupied surface and TWI. There is additionally no correlation between the size of the Thiessen polygons and the TWI.

The study area is marked by a relatively steep decline in rainfall from north to south. In the north, values of 400 millimetres occur while towards the south only 250 millimetres are recorded. Using these observed rainfall distributions the probable settlement pattern was then calculated. For this, a rainfall raster was built into the analysis. Additionally, the investigations included detecting whether there was a relationship between the sum of the settlement area and the rainfall (tab. 1). In all analyses no statistically significant correlation was found. Also, if we divide the settlements into towns and villages, there is no significant correlation with rainfall differences (tab. 1). Finally, there is no significant correlation between the size of the Thiessen polygons and the rainfall patterns (tab. 1).

8. Settlement collapse

It is well-known that significant social changes took place at the end of the third millennium BC in northern Syria. During this period, some settlements reduced in size, while others show evidence for destruction and were deserted.[62] Although the third millennium BC chronology is still in the process of refinement (see also the ARCANE – Associated Regional Chronologies for the Ancient Near East and the Eastern Mediterranean – project), present evidence from excavated North Mesopotamian archaeological sites suggests that the settlement disruptions did not all take place at the same time. Therefore, we need to look for non-catastrophic explanations for the observed patterns.[63] A factor that may have played a role in the observed changes during the end of the third millennium BC is that by 2300 BC a southern Mesopotamian, Akkadian, intervention took place in Syria. Within the Akkadian empire, all southern Mesopotamian city-states were united and a new imperial administration came into rule. The best evidence for Akkadian control over northeastern Syria was found at Tell Brak where an Akkadian administrative building was excavated; it is possibly related to tribute collection from the region and provisions for the troupes. Tell Mozan almost certainly was a powerful Akkadian vassal- or ally-state.[64] The Akkadian kings probably ceded control of the Upper Khabur area during the reign of Sar-kali-sarri (2175-2150 BC), and social turmoil in the north both preceded and followed their departure.[65] Soon afterwards, the kings of Urkiš (Tell Mozan) seized power in northern Mesopotamia and formed the "kingdom of Urkiš and Nawar."[66]

Although the archaeological evidence of wider Mesopotamia suggests a diachronic settlement pattern disruption, survey in the Tell Leilan region within our study area demonstrates a rather synchronous sharp reduction in the number of settlements for the period between 2200 and 1900 BC.[67] This event-like desertion of sites has often been correlated with droughts. The $\delta^{18}O$ isotopic record at Soreq Cave in Israel shows the occurrence of a dry peak between 2150 and 2050 BC after a moist phase from 2550 to 2150 BC.[68] A less obvious dry phase between 2300 and 2100 BC is also visible within the Lake Van isotopic record from southern Turkey.[69] Wilkinson argues that the large-scale EBA societies in northern Syria were particularly prone to longer periods of drought and that the late third millennium urban collapse may have resulted from environmental deterioration caused by urban societies.[70] However, at Tell Mozan, where settlement continued throughout the period between 2200 and 1900 BC and in relation to the newly established kingdom of Urkiš and Nawar, archaeobotanical evidence shows only little evidence for climatic aridification. With continued high proportions of deciduous oak and ubiquities in the

[60] Wilson – Gallant 2000.
[61] Schmidt – Persson 2003.
[62] Akkermans – Schwartz 2003.
[63] see Kuzucuoğlu – Marro 2007.
[64] Akkermans – Schwartz 2003; Sallaberger 2007, 427.
[65] Frayne 1995.
[66] Frayne 1997, 457-464.
[67] Risvet 2005; Staubwasser – Weiss 2006.
[68] Bar-Matthews *et al.* 1998.
[69] Wick *et al.* 2003.
[70] Wilkinson 1997.

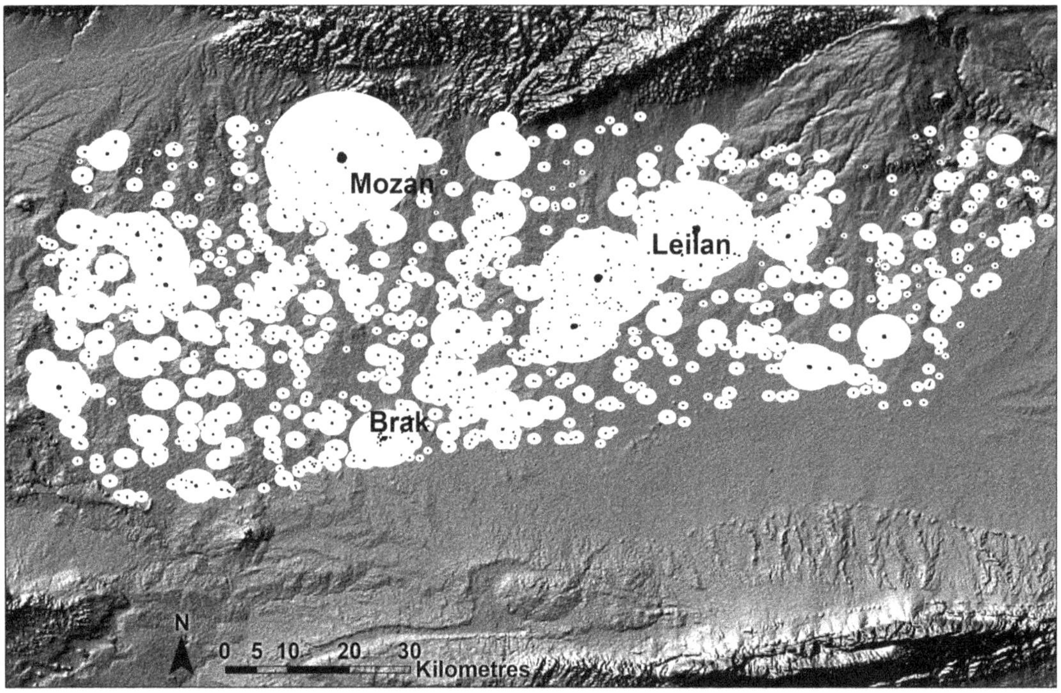

Fig. 7: Estimation of agricultural land necessary to support inhabitants according to proportions established from archaeobotanical data at EBA Tell Mozan and based on the assumption that the people of each settled hectare needed to provide food for 600 persons.

charcoal assemblages, there are hardly any changes visible in the strata between 2250 and 2100 BC. The seed remains from strata dated to 2100-2000 BC even indicate increased soil moisture, which is probably related to the short moisture peak at about this time but indicates no evidence for irrigation. Only from the MBA (2000-1800 BC strata) onwards are reductions in the ubiquity and fragment percentages of deciduous oak, and a shift in the crop assemblage from mainly free-threshing wheat to barley, observed at Mozan. These changes may be tied to an anthropogenic impact on the vegetation related to horticulture and/or general climatic drying.[71] Recent surveys have shown the Tell Leilan region was again densely resettled for 200 years from 1900 BC onwards.[72] During that period the Upper Khabur region was considered to be of strategic importance for the kings of Mari, Ešnunna, Ekall tum, Susa, Babylon, and Aleppo.[73] The 18th century BC texts from Leilan suggest that the countryside was a patchwork of land used for agriculture and pastures.[74]

In order to investigate the impact of tribute on the environment, we needed to increase the number of consumers in the population estimates within our GIS database. Figure 7 was created based on the assumption that each hectare of settlement space needed to provide food for 600 people. If this were in fact the case, the calculations indicate that the resource limits must have been reached in many areas, leaving only between 33.6% and 36.3% of the settlement area uncultivated.

9. Conclusion

Our analysis indicates that during the Bronze Age water, especially streams, but also soil moisture were important factors driving site selection in the Upper Khabur area. The data for the Upper Khabur zone differ somewhat from the eastern Jezira, in that a three-tiered settlement system is not evident within the surroundings of every major centre. Our resource exploitation data further indicated that some larger centres clearly had dependent satellite villages, since they were located within the resource area of these centres. However, along the Wadi Avedji or Wadi Khanzir little overlap of resource exploitation zones was

71 Deckers – Riehl 2007.

72 Risvet 2005; Staubwasser – Weiss 2006.

73 Risvet 2005, 116.

74 Risvet 2005, 132.

visible, which may indicate a degree of self-sufficiency in the villages. The density of occupation along the Wadi Jaghjagh and lack of sufficient cultivatable land within the surroundings of these sites suggests that this river was an important trade route.

Resource calculations that incorporate a tribute that is as high as the necessary harvest to feed the people indicate that resource limits were reached in many areas. Tribute, therefore, may have played a role in the observed settlement collapse at the end of the third millennium BC.

Bibliography

AKKERMANS, P.M.M.G. – SCHWARTZ, G.M.
2003 The archaeology of Syria: from complex hunter-gatherers to early urban societies (c. 16,000–300 BC).

ANGELL, I. – MOORE, R.
1984 On archaeological applications of the Voronoi tessellation, in: S. Laflin (ed.), Computer applications in Archaeology, 9-14.

BAR-MATTHEWS, M. – AYALON, A. – KAUFMAN, A.
1998 Middle to Late Holocene (6,500 yr. period) palaeoclimate in the eastern Mediterranean region from stable isotopic composition of speleothems from Soreq Cave, Israel, in: A.S. Issar – N. Brown (eds.), Water, environment and society in times of climatic change, 203-214.

BEVEN, K.J. – KIRKBY, M.
1979 A physically-based, variable contributing area model of basin hydrology, Hydrological Sciences Bulletin 24, 43-69.

BUCCELLATI, G. – KELLY-BUCCELLATI, M.
1988 The soundings of the first two seasons. Mozan 1. Bibliotheca Mesopotamica 20.

DANKS, P.
1977 Some observations on Medieval and Post-Medieval artifact distributions: A spatial model at the regional scale (macro), in: D.L. Clarke (ed.), Spatial Archaeology, 353-381.

DECKERS, K. – RIEHL, S.
2007 Fluvial deposits from northeastern Syria: Environmental contexts for archaeological sites, Quaternary Research 67, 337-348.
2008 Tells of the Upper Khabur: their hydrological and agrarian contexts, Paléorient 34.2, 173-189.

DRENNAN, R.D. – PETERSON, C.E.
2004 Comparing archaeological settlement systems with rank-size graphs: a measure of shape and statistical confidence, Journal of Archaeological Science 31, 533-549.

GALÁN, M.R.
2005 The Seleucid-Parthian settlement at Tell Beydar and its importance in the context of Hellenistic Upper Mesopotamia, in: M. Lebeau – A. Suleiman (eds.), Tell Beydar/Nabada. An Early Bronze Age city in the Syrian Jezirah: 10 years of research (1992-2002), 38-42.

GOLDHAUSEN, M. – RICCI, A.
2005 Political centralisation in the Syrian Jezira during the 3rd millennium: A case study in settlement hierarchy, Altorientalische Forschungen 32.1, 132-157.

GUICHARD, M.
1994 Au pays de la Dame de Nagar, in: D. Charpin – J.-M. Durand (eds.), Florilegium Marianum II, Recueil d'études à la mémoire de Maurice Birot, Mémoires de NABU, 235-272.

FRAYNE, D.R.
1995 Sargonic and Gutian Periods (2234-2113 BC), Royal Inscriptions of Mesopotamia 2.
1997 Ur III Period (2112-2004 BC), Royal Inscriptions of Mesopotamia 3/2.

HODDER, I. – ORTON, C.
1976 Spatial analysis in archaeology.

KUZUCUOGLU, C. – MARRO, C. (eds.)
2007 Sociétés humaines et changement climatique à la fin du troisième millénaire: une crise a-t-elle eu lieu en Haute Mésopotamie? Actes du colloque de Lyon, 5-8 décembre 2005, Varia Anatolica 19.

KRAMER, C.
1980 Estimating prehistoric populations: an ethnoarchaeological approach, in: M.-T. Barrelet (ed.), L'archéologie de l'Iraq: Perspectives et limites de l'interprétation anthropologique des documents, 315-334.

LEBEAU, M. – SULEIMAN, A.
2005 The city of the Early Jezirah IIIb period, in: M. Lebeau – A. Suleiman (eds.), Tell Beydar/Nabada. An Early Bronze Age city in the Syrian Jezirah: 10 years of research (1992-2002), 13-26.

LYONNET, B.
1990 Prospection archéologiques du site de Tell Mohammed Diyab, Cahiers de Nabu 1, 71-115.
1996 Settlement pattern in the Upper Khabur (N.E. Syria) from the Achaemenids to the Abbasid Period: Methods and preliminary results from a survey, in: K. Bartl – S.R. Hauser (eds.), Continuity and change in Northern Mesopotamia from the Hellenistic to the Early Islamic Period, 349-354.
1997 Questions sur l´origine des porteurs de pots en Haute-Mésopotamie, du VIe au milieu du IIe millénaire, in: M.T. Barrelet – D. Charpin (eds.), Recueil d'études à la mémoire de Marie-Thérèse Barrelet, Florilegium marianum 3, 133-144.
1998 La prospection archéologique de la partie occidentale du Haut-Khabur (Syrie du Nord-Est): méthodes, résultats et questions autour de l´occupation aux IIIe et IIe millénaires av. n.e., in: J.M. Durand (ed.), Mari et les hourites (Amurru 1), 363-376.

LYONNET, B. (ed.)
2000 Prospection archéologique du Haut-Khabur Occidental (Syrie du N.E.), Vol. 1.

MATTHEWS, R. (ed.)
2003 Excavations at Tell Brak. Vol. 4: Exploring an Upper Mesopotamian regional centre, 1994-1996.

MATTHEWS, R.
2003 A chiefdom of the northern plains, in: R. Matthews (ed.), Excavations at Tell Brak. Vol. 4: Exploring an Upper Mesopotamian regional centre, 1994-1996, 25-51.

MEIJER, D.J.W.
1986 A survey in Northeastern Syria.

MILANO, L – ROVA, E. – STÉNUIT, M.E.
2005 Tell Beydar: The early phases, in: M. Lebeau – A. Suleiman (eds.), Tell Beydar/Nabada. An Early Bronze Age city in the Syrian Jezirah: 10 years of research (1992-2002), 4-12.

PFÄLZNER, P. – WISSING, A.
2004 Urbanismus in der Unterstadt von Urkes, Mitteilungen der Deutschen Orient Gesellschaft zu Berlin 136, 41-86.

RISVET, L.
2005 Settlement, economy, and society in the Tell Leilan region, Syria, 3000-1000 BC, Cambridge: King's College, unpublished PhD dissertation.
2008 Legal and archaeological territories of the second millennium BC in northern Mesopotamia, Antiquity 82, 585-599.

RÖLLIG, W.
1984 Preliminary remarks on the Middle Assyrian archive from Tall Seh-Hamad/Dur-Katlimmu, Les Annales Archéologiques Arabes Syriennes 34, 189-194.

SALLABERGER, W.
2005 The third millennium cuneiform texts from Beydar, in: M. Lebeau – A. Suleiman (eds.), Tell Beydar/Nabada. An Early Bronze Age city in the Syrian Jezirah: 10 years of research (1992-2002), 27-32.
2007 From urban culture to nomadism: A history of Upper Mesopotamia in the late third millennium, in: C. Kuzucuoğlu – C. Marro (eds.), Sociétés humaines et changement climatique à la fin du troisième millénaire: une crise a-t-elle eu lieu en Haute Mésopotamie? Actes du colloque de Lyon, 5-8 décembre 2005, Varia Anatolica 19, 417-437.

SCHMIDT, F. – PERRSON, A.
2003 Comparison of DEM Data Capture and Topographic Wetness Indices, Precision Agriculture 4-2, 179-192.

STAUBWASSER, M. – WEISS, H.
2006 Introduction: Holocene climate and cultural evolution in Late Prehistoric-Early Historic West Asia, Quaternary Research 66, 372-387.

STEELE, C. – MCDONALD, H. – MATTHEWS, R. – BLACK, J.
2003 Impact of the Empire. Later third-millennium investigations: the Late Early Dynastic and Akkadian periods, in: R. Matthews (ed.), Excavations at Tell Brak. Vol. 4: Exploring an Upper Mesopotamian regional centre, 1994-1996, 193-269.

STEIN, G. – WATTENMAKER, P.
1990 The 1987 Tell Leilan Regional Survey: Preliminary report, in: N. Miller (ed.), Economy and settlement in the Near East. Analyses of ancient sites and materials: 5-18, MASCA Research Papers in Science and Archaeology Supplement 7, 9-18.

UR, J.
2003 CORONA satellite photography and ancient road networks: A northern Mesopotamian case study, Antiquity 77, 102-115.
In press. Urban form at Tell Brak across three millennia, in: H. Crawford – A. McMahon (eds.), Preludes to urbanism: Studies in the Late Chalcolithic of Mesopotamia in honour of Joan Oates, British Archaeological Reports, International Series.

UR, J. – KARSGAARD, P. – OATES, J.
2007 Early urban development in the Near East, Science 317, 1188.
In press. The spatial dimensions of Early Mesopotamian Urbanism: The Tell Brak suburban survey, 2003-2006, Iraq 73.

UR, J. – WILKINSON, T.J.
2008 Settlement and economic landscapes of Tell Beydar and its hinterland, Subartu 77, 305-327.

VAN DER STEDE, V.
2005 The Akkadian occupation, in: M. Lebeau – A. Suleiman (eds.), Tell Beydar/Nabada. An Early Bronze Age city in the Syrian Jezirah: 10 years of research (1992-2002), 32-36.

VAN LIERE, W.J. – LAUFFRAY, J.
1954 Nouvelle prospection archéologique dans la Haute Jezireh syrienne, Les Annales Archéologiques de Syrie 4-5, 129-148.

WÄFLER, M.
1990 The Excavations at Tell Hamidi, in: S. Eichler – M. Wäfler (eds.), Tall al-Hamidiya 2, Orbis Biblicus et Orientalis V. 6, 219-228.
1995 Kahat, Tâdum, und Ilansura, NABU 31, 25-27.
2001 Tell al-Hamidiya 3: Zur historischen Geographie von Idamaras zur Zeit der Archive von Mari und Shubat-enlil/Shehna, Orbis Biblicus et Orientalis Series Archaeologica 21.

WÄFLER, M. – WARBURTON, D.A. – EICHLER, S. – STEUDLER, D.
1985 Tall al-Hamidiya 1, Vorbericht 1984, Orbis Biblicus et Orientalis, V. 4.

WEISS, H. – COURTY, M.-A. – WETTERSTROM, W. – GUICHARD, F. – SENIOR, L. – CURNOW, A.
1993 The genesis and collapse of third millennium north Mesopotamian civilization, Science 261, 995-1004.

WICK, L. – LEMCKE, G. – STURM, M.
2003 Evidence of Lateglacial and Holocene climatic change and human impact in eastern Anatolia: High-resolution pollen, charcoal, isotopic and geochemical records from the laminated sediments of Lake Van, Turkey, The Holocene 13.5, 665-675.

WILKINSON, T.J.
1994 The structure and dynamics of dry-farming states in Upper Mesopotamia, Current Anthropology 35.5, 483-520.
1997 Environmental Fluctuations, Agricultural Production and Collapse: A View from Bronze Age Upper Mesopotamia, in: H.N. Dalfes – G. Kukla – H. Weiss (eds.), Third Millennium BC Climate Change and Old World Collapse, NATO ASI Series I: Global Environmental Change 49, 67-106.
2000 Archaeological survey of the Tell Beydar Region, Syria, 1997: A preliminary report, in: K. Van Lerberghe – G. Voet (eds.), Tell Beydar Environmental and Technical Studies, Subartu 6, 1-37.
2002 The settlement transition of the second millennium BC in the western Khabur, in: L. Al-Gailani Werr – C. Curtis – H. Martin – A. McMahon – J. Oates – J. Reade (eds.), Of pots and plans: Papers on the archaeology and history of Mesopotamia, 361-372.

2003 Archaeological landscapes of the Near East.

WILKINSON, T.J. – BARBANES, E.

2000 Settlement patterns in the Syrian Jazira during the Iron Age, in: G. Bunnens (ed.), Essays on Syria in the Iron Age, Ancient Near Eastern Studies Supplement 7, 397-422.

WILKINSON, T.J. – TUCKER, D.J.
1995 Settlement development in the North Jazira, Iraq. A study of the archaeological landscape.

WILKINSON, T.J. – FRENCH, C.A.I. – MATTHEWS, W. – OATES, J.
2001 Chapter 1. Geoarchaeology, Landscape and the Region, in: D. Oates – J. Oates – H. McDonald (eds.), Excavations at Tell Brak V. 2: Nagar in the third millennium BC, 1-14.

WILKINSON, T.J – CHRISTIANSEN, J.H. – UR, J. – WIDELL, M. – ALTAWEEL, M.
2007 Urbanization within a dynamic environment: Modeling Bronze Age communities in Upper Mesopotamia, American Anthropologist 109.1, 52-68.

WILSON, J.P. – GALLANT, J.C.
2000 Terrain Analysis. Principles and applications.

3. Vegetation development in relation to human occupation and climatic change in the Middle Euphrates and Upper Jazirah (Syria/Turkey) during the Bronze Age

Katleen Deckers and Hugues Pessin

1. Introduction

The Middle Euphrates and Upper Jazirah show a strong gradient in rainfall from north to south, with values of more than 450 mm in the north and as little as 150 mm in the south. The 250 mm isohyet marks the conventional dry farming border, south of which farming becomes difficult without the use of irrigation.

Today, within the study zone, woody vegetation mainly occurs along the Euphrates River. It consists of *Populus euphratica* (Euphrates poplar), *Salix acmophylla* (weeping willow), and *Tamarix tetendra* or *T. smyrnensis* (tamarisk). Khabur riparian forest vegetation is hardly existent. Moreover, along the Wadi Jarrah in the Upper Khabur, no riverine woody vegetation can be found because the river has been dammed and no longer contains perennial water. Hardly any woody vegetation can be found away from the rivers in Syria. However, the nearest deciduous oak woodland can be found circa twelve kilometres north of the border at the same elevation as Tell Mozan, about sixty kilometres north of the border near the Balikh and roughly twenty-five kilometres north of the Syrian border at the Middle Euphrates (fig. 1). Hillman[1] reconstructed the potential vegetation for this area – to be exact, the vegetation under present day climatic conditions in the absence of human impact. His map is based on studies of today's remaining vegetation, rainfall levels, insulation and temperature levels, altitude, topography, aspect effects, solid geology, soil types, patterns of concentration of less ephemeral wadis as indicated on topographical maps, and accounts of early travellers. Hillman's data collection also incorporated relative densities of settled villages and farmsteads on large-scale topographical maps, which predate the widespread availability of irrigation pumps. He concluded the region under study should have been somewhat greener, specifically with open deciduous park woodland in the north and *Pistacia* (pistachio)-*Amygdalus* (almond) woodland steppe toward the south. Hillman also deduced the riverine vegetation would have been more lush and diverse (fig. 1).

Until recently, little was known about the vegetation history of the Middle Euphrates and Upper Jazirah during the Bronze Age due to poor pollen preservation.[2] Anthracology, therefore, plays a key role in reconstructing the vegetation within this zone. Although other factors may have played a role in the proportion of woody taxa at a site, it is often assumed that, especially in the case of firewood, they approximate the relative abundance of taxa in the vegetation of the area. Based on this assumption, the amount of a charcoal taxon within samples and its ubiquity throughout the samples from a site are used as indicators of the relative abundance in the local vegetation.[3] Since it has been demonstrated that weighing samples provides comparable results to counting,[4] and because fragments within this area were counted in previous studies, we report only results based on fragment counts. Several anthracological publications demonstrated the potential of anthracology within this region.[5] In this chapter, we provide an overview of vegetation changes based on anthracological analysis within this area throughout the Bronze Age (3000-1200 BC) and evaluate causes for the observed changes.

The Bronze Age within this region covers a period of roughly 1800 years. At the start of the Early Bronze Age (EBA) societies were quite rural. Urbanisation developed from the mid third millennium BC onwards and an explosion in settlement took place.[6] This was a period of city states. A hierarchical settlement pattern came into existence, with fortified cities and dependent satellite communities. Together with this urban revolution, economic specialisation and increased large-scale trade took place. Ceramics

---

1 in Moore *et al.* 2000.

2 Bottema 1989.

3 Smart – Hoffman 1988.

4 Miller 1985; Chabal 1991.

5 Willcox 2002; Pessin 2004; 2007; Deckers 2005; Deckers – Riehl 2007b.

6 Akkermans – Schwartz 2003.

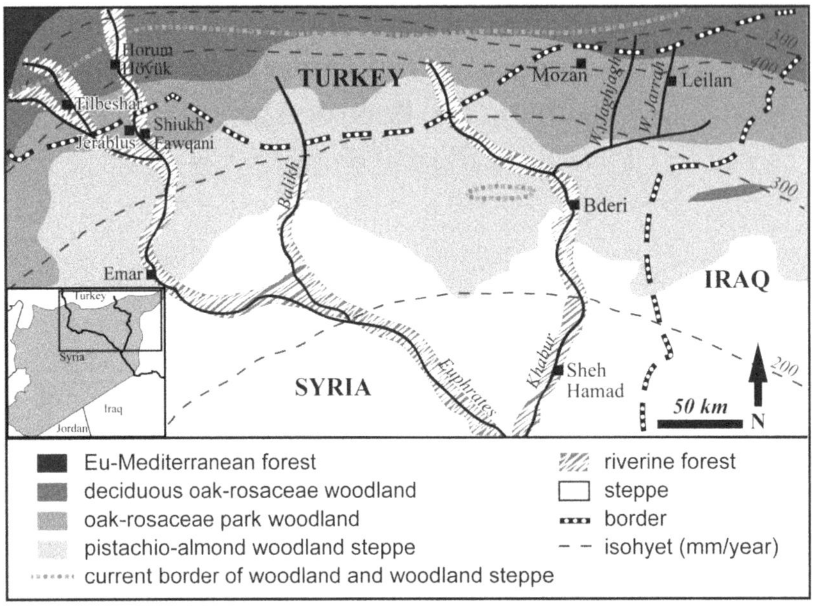

Fig. 1: Potential vegetation map with location of sites (after Hillman 2000). The present day woodland and woodland steppe borders were delineated based on satellite image analysis and field observations.

became mass products, metals were more intensively used, and an increased specialisation within the agrarian and pastoral sector occurred. Towards the end of the EBA some settlement disruptions took place. Subsequently, Mari developed into a powerful state and controlled the area under investigation (ca. 2000-1760 BC). A number of epigraphic sources inform us about the export of wine, wood, and olive oil from other Syrian towns (e.g. Karkami and Emar) to southern Mesopotamia via Mari.[7] It is of interest here that King Yahdun-Lim claims he was the first King to reach the western sea and cut cedars and pines from the mountains.[8] [9] At about 1760 BC, Hammurabi of Babylon destroyed Mari, bringing its years of domination to an end. Additionally, around 1595 BC the Hittite invasion changed the political climate. During the Late Bronze Age (LBA), the region suffered domination from both Egypt and Anatolia (Hittites), endured a major Hurrian dynasty in the Khabur region (Mitanni), and was also under nomadic pressure from the Arameans. At the end of the LBA, much of the region under consideration came under the control of Assyria.[10]

In order to evaluate the impact of ancient societies in the area, we must take into account the intensity of the settlements for each of the periods investigated. This is possible thanks to data gleaned from the intensive archaeological surveys that have occurred in northern Syria/southern Turkey. The picture differs regionally. For the northern Middle Euphrates region, it was found that from the EBA to the LBA an increase in the inhabited area took place. Beginning at roughly forty hectares in the EBA, the occupied land grew to 110 ha in the Middle Bronze Age (MBA), and reached 150 ha in the LBA.[11] On the other hand, Wilkinson[12] found that in the southern Middle Euphrates there was a general decrease in the number of sites and area occupied over time. This was clearly noted in the Sweyhat and Upper Lake Assad region throughout the Bronze Age. In the western Upper Khabur, the main occupation phase for many large sites occurred in the EBA, with only a few small sites being occupied in the MBA and LBA. In the eastern Upper Khabur, surveys indicated a steady increase in settled area during the EBA and a sharp decrease at the end of the EBA; but the main occupation phase was at the beginning of the MBA.[13]

---

[7] Hempel 2003.

[8] The first king who claimed to have reached the Mediterranean Sea is Sargon of Akkad (Frayne 1997) and the first expedition that ever mentioned the forest of cedars appeared in the Epic of Gilgamesh (George 2003).

[9] Taraqji 1997.

[10] Chavalas 1992.

[11] Algaze *et al.* 1994.

[12] Wilkinson 2004.

[13] Risvet 2005.

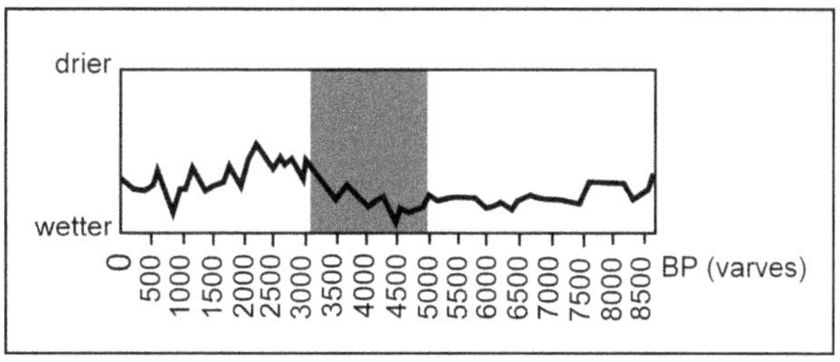

Fig. 2: Lake Van paleoclimatic proxy data based on stable oxygen isotopes (after Lemcke – Sturm 1997).

In addition to correlating our data with the settlement patterns, we must also compare vegetation changes with the climatic evolution known for this region and period. For this, the stable oxygen isotopic curve of Lake Van in southeastern Turkey is an important climatic proxy.[14] Although fluctuations occurred, a trend of climatic drying can generally be observed from the EBA to the LBA, with climatic conditions being moister today than during the LBA. However, the MBA was as dry as today's levels (fig. 2). A similar pattern of climatic drying through the Bronze Age has also been found for Soreq Cave in Israel.[15]

2. The samples

This overview of vegetation changes during the Bronze Age is based on charcoals from nine archaeological sites (fig. 3 and tab. 1). Most of the sites discussed have a riverine location, which is not always obvious at present. This is particularly the case for Tell Leilan, where the Jarrah has been dammed and runs dry. Tell Mozan is the only site that was not located along a permanent stream.[16]

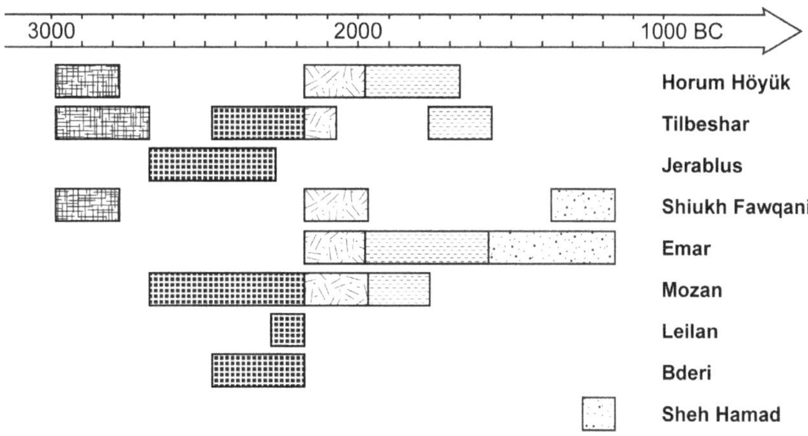

Fig. 3: List of sites with periods covered by anthracological data.

Figure 3 summarises the periods covered by the anthracological data from these sites. Table 1 depicts the number of samples and fragments for each site and shows details of fragment counts and ubiquity scores for all the sites and periods. In total, this summary contains more than 50,000 fragments from 385 samples, most of which were identified during our study. Published data from Tell Sheh Hamad[17] and Tell Bderi[18] have been included within this overview as well. Most data are derived from floated samples. From Tell Mozan, flotation samples as well as large sediment samples from charcoal concentrations have been investigated and should be representative for the local vegetation. A range of contexts have also been analysed,[19] mostly deriving from secondary fill deposits. Charcoal samples from a few abandonment layers, hearths, and

[14] Lemcke – Sturm 1997; Wick *et al.* 2003.

[15] Bar-Matthews – Kaufmann 1998.

[16] see also chapter 4 this volume.

[17] Frey *et al.* 1991.

[18] Engel 1993.

[19] see e.g. Deckers 2005 for details of Emar.

Table 1: Sites with the number of samples and fragments analysed.

Site	HORUM HÖYÜK						TILBESHAR								JERABLUS				SHIUKH FAWQANI						EMAR					
Period (BC)	3000-2800		2200-2000		2000-1700		3000-2700		2500-2200		2200-2000		1800-1600		3000-2700		2700-2200		3000-2800		2200-2000		1400-1200		2200		2000-1600		1600-1200	
Frequency & ubiquity	AF	U	AF	U	AF	U	AF	U	AF	U	AF	U	AF	U	AF	U	AF	U	AF	U	AF	U	AF	U	AF	U	AF	U	AF	U
Acer sp.					1	1											1	1												
Alnus sp.					3	2			3	2	5	1	2	1					8	3	11	3	68	4	6	2				
Amygdalus sp.			3	2	10	2	12	5	26	6	4	1	3	2			7	2												
Buxux sp.																														
Cedrus sp.																	14	1												
Chenopodiaceae					2	1	2	1			1	1													7	2			1	1
Clematis sp.																	1	1												
Conifer sp.																									1	1			9	1
Cupressus sp.																									2	2				
Eleagnus angustifolia																														
Ficus carica					25	1							1	1	8	1	26	6							2	2			8	3
Fabaceae																	1	1	9	2	2	1								
Fraxinus sp.	3	2	8	2	227	2	305	10	100	7	32	2	33	4			105	7	95	11	2	1	69	2	4	2			19	1
Juglans sp.															70	1														
Juniperus sp.					3	2	2	1	18	3	3	1					4	2	2	2			3	1						
Lycium sp.																									20	5	39	3	18	6
Monocotyledon									2	1					8	3	29	5							317	3	6	2	12	3
Morus alba																														
Olea europea			14	3	451	6	4	1	73	7	35	2	29	3			14	2							3	2				
Paliurus spina-christi					1	1	2	1											2	2	2	1								
Phoenix sp.																	45	3												
Phragmites autralis			1	1	1	1	29	9	11	5	2	1	1	1					59	9	6	2	67	8	1	1			38	1
Pinus brutia/halepensis			1	1	52	3	9	3	41	5	45	2	14	2							11	2	80	4	1	1			14	3
Pinus nigra/sylvestris					2	1	6	3					1	1			22	3	6	4										
Pistacia atlantica	7	1	6	1	9	3	12	5	29	7	6	1	31	3			16	2	3	2										
Pistacia sp.																	5	2							5	3	19	1		
Platanus orientalis			2	1	87	2	7	2	10	4	2	1	9	2			4	1	28	5	6	1	16	2	8	4				
Platanus sp./*Fagus* sp.																	3	2												
Pomoideae	1	1			23	2	36	8	22	7	5	1	4	1					6	3			2	1			1	1		
Prosopis farcta																														
Prunus sp.							143	4	10	4	1	1							6	3									4	1
Quercus deciduous	329	7	138	3	1783	7	856	10	669	8	292	2	405	4	1	1	95	6	59	11	1	1	159	3	8	1				
Quercus evergreen																													1	1
Quercus sp.																	7	2							3	2				
Rhamnus sp.									1	1			1	1																
Salix sp./*Populus* sp.	11	3	4	2	129	6	59	8	13	5	8	1	7	4	810	6	1510	7	284	16	283	3	652	8	17504	11	1632	8	5275	17
Tamarix sp.	1	1			12	2			4	3	1	1	3	1	1625	3	158	5	217	14	5	1	3	1	673	9	393	5	549	10
Ulmus sp.	1	1			27	4	48	7	3	1			1	1					13	1	58	3			2	2	1	1	612	3
Vitis vinifera					7	3	6	4	20	4	8	1	6	2			10	1											1	1
Ziziphus sp.																									2	1				
Zygophyllum sp.																														
Sum	**353**	**7**	**177**	**4**	**2855**	**9**	**1538**	**11**	**1055**	**8**	**450**	**2**	**551**	**4**	**2522**	**6**	**2077**	**9**	**798**	**17**	**387**	**3**	**1119**	**8**	**18551**	**11**	**2091**	**8**	**6543**	**20**

Site	MOZAN						LEILAN		BDERI	SHEH HAMAD	
Period (BC)	2700-2200		2200-2000		2000-1800		2300-2200		2500-2200	13th century	
Frequency & ubiquity	AF	U	AF	U	AF	U	AF	U	AF	AF	U
Acer sp.									1		
Alnus sp.											
Amygdalus sp.							4				
Buxux sp.										15	1
Cedrus sp.			2	1							
Chenopodiaceae	3	2	49	1					14	1	1
Clematis sp.											
Conifer sp.			4	1							
Cupressus sp.	1	1	1	1							
Eleagnus angustifolia									2		
Ficus carica	74	6	2	2	13	1			19		
Fabaceae											
Fraxinus sp.	273	15	40	10	14	2	155	5	170		
Juglans sp.											
Juniperus sp.	3	1	3	2							
Lycium sp.									4	2	1
Monocotyledon	18	9	585	4	49	2	4	3	10		
Morus alba									43		
Olea europea					474	5			1		
Paliurus spina-christi											
Phoenix sp.									5		
Phragmites autralis									57		
Pinus brutia/halepensis	4	1	6	2	85	1					
Pinus nigra/sylvestris											
Pistacia atlantica	20	7	0	0	1	1					
Pistacia sp.							27	1	4		
Platanus orientalis	37	7	39	8	43	2			5		
Platanus sp./*Fagus* sp.										15	2
Pomoideae	25	6	22	5	35	2	6	1	12	5	2
Prosopis farcta										15	2
Prunus sp.	1	1	2	1							
Quercus deciduous	3608	54	2411	51	332	26	79	6	83		
Quercus evergreen											
Quercus sp.	135	18	268	12			1	1			
Rhamnus sp.									1		
Salix sp./*Populus* sp.	206	28	117	17	125	9	83	8	1040	87	9
Tamarix sp.	1	1							286	158	8
Ulmus sp.	64	5	27	3	16	2	47	4	37	23	2
Vitis vinifera			2	1	14	2	1	1	6		
Ziziphus sp.											
Zygophyllum sp.									1		
Sum	4101	59	2894	60	651	43	407	9	1801	321	17

construction wood have also been included within the overview here. The latter context is represented only in very few cases.

3. Interpretation of the wood charcoal assemblages in relation to past habitats

Considering the large number of data, fragment percentages were grouped into five vegetation categories, specifically, open park woodland, riverine forest, imported taxa, cultivated taxa, and a non-specifiable class.

Deciduous *Quercus* (oak), *Pistacia, Amygdalus,* Pomoideae, *Prunus, Juniperus* (juniper), *Rhamnus* (buckthorn), *Lycium* (boxthorn), *Paliurus*, and *Ziziphus* were interpreted as belonging to open park woodland or woodland steppe.[20] Within this region, the main woody taxon of the open park woodland would consist of *Quercus*, whereas *Pistacia* would indicate woodland steppe.[21] Overall, within this paper, references to open park woodland imply a somewhat lusher woody vegetation than the one assumed for woodland steppe. However, using charcoal identifications it remains difficult to draw a clear border between park woodland and woodland steppe since they form a continuum. There is also a considerable species overlap between the two vegetation units and it is difficult to reconstruct the density of the vegetation.

Salix/Populus, Tamarix, Fraxinus (ash), *Phragmites* (reed), *Alnus* (alder), *Clematis, Platanus* (plane), *Eleagnus Angustifolia* (oleaster), and *Ulmus* (elm) were classified as taxa of the riverine forest.[22] There exist, however, some 24th century BC texts from southern Mesopotamia that indicate (if the words were translated correctly) that besides occurring naturally in the riverine woodland, *Populus* and *Tamarix* were also planted.[23]

Furthermore, it is impossible to distinguish wild from cultivated fruit trees by means of basic wood anatomy. Although the present day natural distribution of *Vitis* (grapevine) and *Ficus* (fig) indicates some may have been growing wild in the local surroundings,[24] we classified them among the cultivated taxa. Seeds of these fruits were also found at the investigated sites[25] and some texts (especially for the LBA) also confirm the presence of vineyards.[26] In another instance, *Olea* (olive) and *Phoenix* (palm tree) charcoal were found outside their natural distribution areas, suggesting they were cultivated. Further signs of local cultivation were seen at Horum Höyük, Tilbeshar, and Emar, where olive stones were found.[27] Nevertheless, this was not the case in MBA Mozan,[28] which may be due to the fact that olive oil production took place in pre-assigned areas with garbage deposition nearby.

Based on the present geographical distribution of trees, the following were counted among the imported taxa: *Pinus nigra/sylvestris* (black pine/Scots pine), *Cedrus* (cedar), *Cupressus* (cypress), *Buxus* (boxwood), and indeterminate conifer forms.[29] The real situation, however, may have been more complex, as suggested by a text from Mari concerning boxwood. Specifically, saplings of this tree species were imported to be planted.[30]

In the unspecifiable group, we listed *Pinus brutia/halepensis* (Calabrian pine/Aleppo pine), evergreen oak, *Morus* (mulberry), *Prosopis*, monocotyledons, *Zygophyllum* (bean-caper), Chenopodiaceae, Fabaceae, *Juglans* (walnut), and *Acer* (maple).[31] The imported status of the evergreen oak is somewhat debatable. This tree occurs today in scrub varieties within the Mediterranean zone and its wood, generally, only reaches small dimensions in growth; therefore, it is unlikely to have been imported. Additionally, Calabrian pine could have been present here in the study area as a coloniser after fire clearance, or because Mediterranean vegetation may have extended to this area. Chenopodiaceae are mainly drought indicators but may equally have occurred in the woodland steppe as in the steppe. *Morus* was possibly cultivated, though it more likely arose naturally. Furthermore, *Juglans* may have been either cultivated, or imported. *Prosopis* occurs on alluvial soils, among crops and on saline grounds and river banks, and may be an indicator of browsing. Its

---

[20] Zohary 1973.

[21] Moore *et al.* 2000.

[22] Zohary 1973.

[23] Powell 1992.

[24] see Zohary – Hopf, 2000.

[25] e.g. Deckers – Riehl 2007b; Riehl 2010; Willcox 2010.

[26] Beckman 1996; Westenholz 2000.

[27] Deckers – Riehl 2007b; Riehl 2010; Willcox 2010.

[28] Deckers – Riehl 2007b.

[29] Zohary 1973.

[30] Michel 1996, 389.

[31] Michel 1996, 389.

charcoal was only found at Bderi. *Zygophyllum* is a typical dwarf scrub that occurs in deserts and steppes, and similarly may be an indicator of human disturbance. The Fabaceae were also classified into the unspecifiable group since it is a large family and their species grow in many different environments. Finally, *Acer* could likewise be an indicator of riverine forest, as well as of woodland steppe or Mediterranean-type vegetation. Apart from *Pinus brutia/halepensis*, all these taxa occurred in rather small quantities and few samples.

4. Criteria for the evaluation of climatic and human impact on the vegetation

In order to evaluate climatic and human impact on the vegetation, ubiquities and ubiquity ratios of some taxa will be examined in more detail.

Upon drying climatic conditions, the southern part of the oak park woodland/northern border of the *Pistacia-Amygdalus* woodland steppe would have retreated northwards towards wetter locations.[32] Therefore, the *Pistacia*-deciduous *Quercus* ubiquity ratio is a proxy for climatic change. It is of course understood that a diversity of species of the *Pistacia-Amygdalus* steppe could also occur within the oak park woodland, so their presence/absence does not demonstrate the location of the border between the two.

Additionally, fruit tree charcoal ubiquity will be used as a measure of the importance of cultivation in addition to fragment proportions. Fruit tree wood remains primarily came into the archaeological record as the result of pruning, and fruit tree wood was mainly used as firewood.[33] Moreover, *Vitis* wood does not generally attain great dimensions. Therefore, the ubiquity of fruit trees provides a more appropriate picture of the impact of horticulture than fragment count percentages.

Another indicator for former human impact on the vegetation is the *Populus/Salix* to *Tamarix* ratio, which may indicate the impact of damming and irrigation on the vegetation. This is because under a pulse flood regime, most studies show that *Populus/Salix* outcompete *Tamarix*. However, when rivers are managed and dammed, *Tamarix* tends to expand at the expense of *Populus/Salix*. Increased proportions of *Tamarix* relative to mesophytic trees like *Populus/Salix* may therefore indicate human impact on the river systems; in example, damming and irrigation resulting in decreased pulse flood regimes.[34] Climatically-caused reductions in flooding may have had similar effects.

In addition to the *Populus/Salix* to *Tamarix* ratio, the general diversity of the riverine vegetation can also be used as a proxy for human impact. Since the Euphrates and Khabur were perennial and dependent on exogenous water from the eastern Anatolian Mountains, their riparian vegetation, especially the diversity, would not have been very sensitive to reduced rainfall. Therefore, a reduced diversity within the charcoal samples may be due to human impact on the riverine vegetation. The opposite, however, can also be argued – that a low diversity of riverine species within the samples may actually be indicative of the availability of plenty of woody resources and, as a result, people were able to select those taxa they preferred most. If all riverine taxa were amply available and gaining good fuel wood was a main strategy[35] at the sites, one would, for example, expect *Fraxinus*, *Ulmus*, *Platanus*, and *Fagus* to be the dominant riverine species within the samples, while *Populus*, *Salix* and *Alnus* would be poorly represented. It will be demonstrated below that this is not the case.

In order to understand better wood exploitation strategies within the park woodland vegetation, diameter measurements were undertaken on *Quercus* charcoal.[36] The diameter was measured with a digital Keyence microscope and was then classified into groups.

5. Results and discussion

5.1 Woodland

The proportions of the vegetation categories (fig. 4) demonstrate the presence of a more southward distribution of park woodland during the Bronze Age in southern Turkey and northern Syria. This is consistent with Hillman's[37] reconstruction of the potential vegetation (fig. 1). In the north, the open park woodland

---

[32] see Moore *et al.* 2000.

[33] Gale – Cutler 2000, 173.

[34] Glenn – Nagler 2005.

[35] In Emar, for instance, it is obvious that the main riverine taxa present (*Populus/Salix* and *Tamarix*) were regularly used as fuel. This is obvious by their presence in oven contexts, but also by the high ubiquity of small twigs of these species within the charcoal assemblage (Deckers 2010).

[36] see e.g. Marguerie – Hunot 2007.

[37] Hillman 2000 in Moore *et al.* 2000.

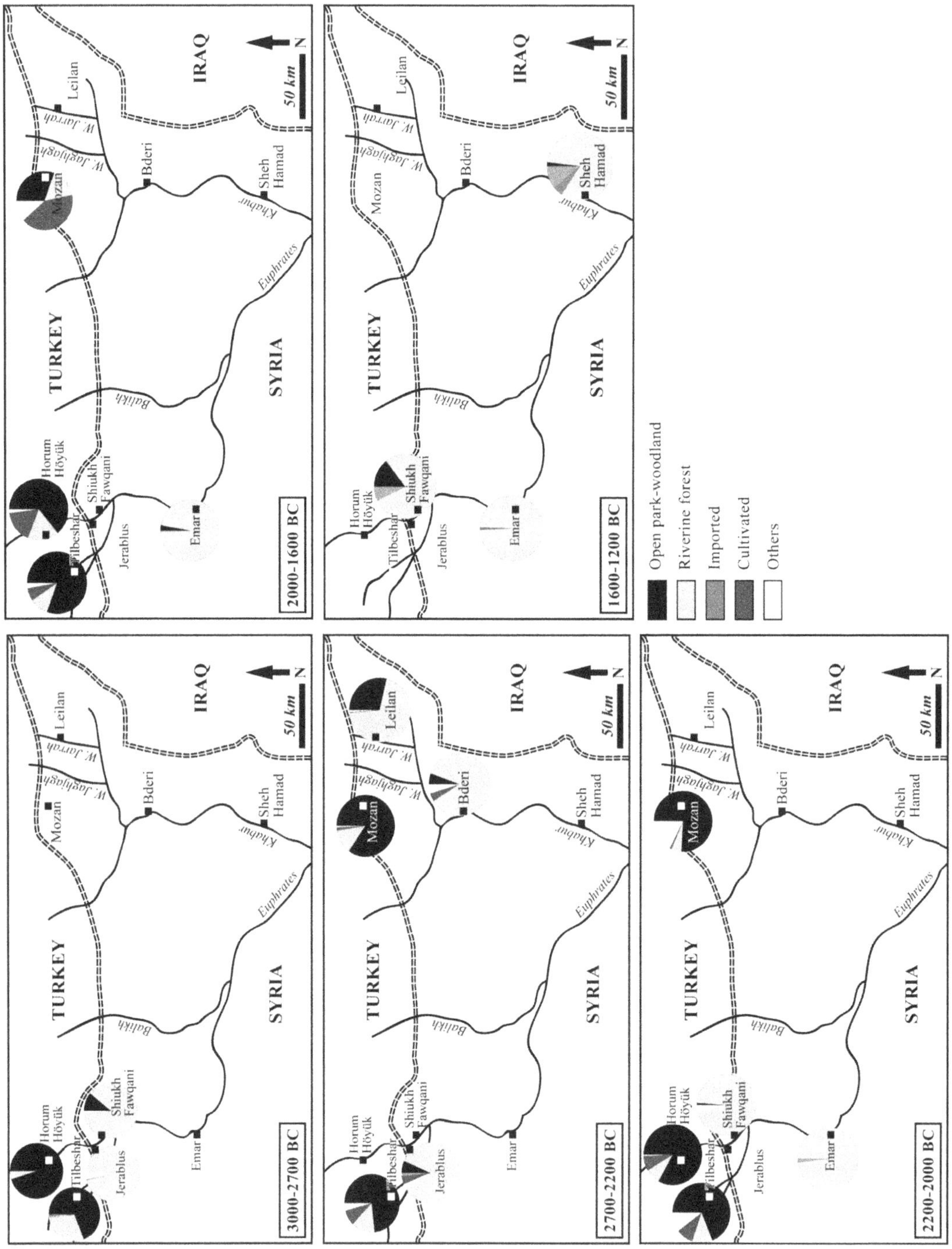

Fig. 4: Pie chart summaries of charcoal fragment proportions for each period.

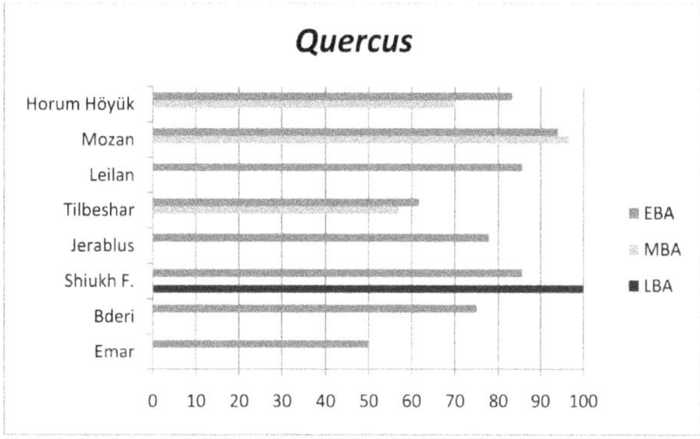

Fig. 5: *Quercus* ubiquity percentages in proportion to *Pistacia*. Sites are listed from north to south (higher rainfall in the north).

was dominated by deciduous oak, while *Pistacia* dominated towards the south (fig. 5). However, woodland remains in the latter area were rather scarce in the Late EBA in comparison to the Early Holocene, which may be indicative of early deforestation or a result of the onset of drier climatic conditions. Interestingly, in most sites, an increase in the ubiquity percentages of *Pistacia* relative to oak can be observed through time; this may relate to climatic drying and retreat of the *Pistacia* woodland steppe and open oak park woodland toward the north (fig. 5). Exceptions are found in Tell Mozan and Tell Shiukh Fawqani.

That these woodland taxa were locally present is indicated by the small twigs of oak found at Tell Mozan, Horum Höyük, and Tilbeshar, and the acorns found at Tell Mozan. A more detailed look at preliminary results of diameter measurements of deciduous *Quercus* fragments from Jerablus, Leilan, and Mozan, indicates that only small branches were used, and that the whole tree was not cut down, but instead coppicing was practiced (tab. 2). More precisely, of the thirty-seven measured diameters of the charcoal fragments, all were smaller than twelve centimetres. Oaks lent themselves well for coppicing[38] and, consequently, for consistently providing firewood to the region's people in a manner that left the trees undamaged. Thalen,[39] who undertook ecological studies in Iraq, noted that the coppicing of trees and shrubs does not have adverse consequences for the vegetation. It is only when this practice occurs excessively in the same locale that a detrimental effect arises.

Table 2: Diameter measurements for deciduous Quercus *charcoal fragments from Jerablus, Mozan, and Leilan.*

	Dates	0-2 cm	2-3 cm	3-5 cm	5-10 cm	10-15 cm	>15 cm
Jerablus	3000-2800 BC	-	1	1	-	-	-
	2700-2300 BC	3	-	4	4	1	-
Leilan	2300-2200 BC	-	-	1	1	-	-
Mozan	2600-2400 BC	1	1	3	-	-	-
	2300-2100 BC	1	-	4	-	-	-
	2100-2000 BC	1	2	1	3	-	-
	2000-1800 BC	1	2	1	-	-	-

The 2200-2000 BC reduction in fragment percentages of oak at the sites could be a result of the dry climatic peak between about 2300 and 2100 BC, as found at Lake Van,[40] which caused a reduction of park woodland vegetation. The general trend, nonetheless, of increased proportions of park woodland taxa at the expense of riverine taxa at three of the four riverine sites where diachronic data is available (Shiukh Fawqani, Tilbeshar, and Jerablus), suggests that riverine environments were exploited and cleared first and that only in subsequent phases was there an increased exploitation of the open park woodland zone. This pattern may have been driven by a strongly augmented population size in the northern Middle Euphrates region from the EBA to LBA.[41]

[38] Gale – Cutler 2000.

[39] 1979.

[40] Wick *et al.* 2003.

[41] Algaze 1994.

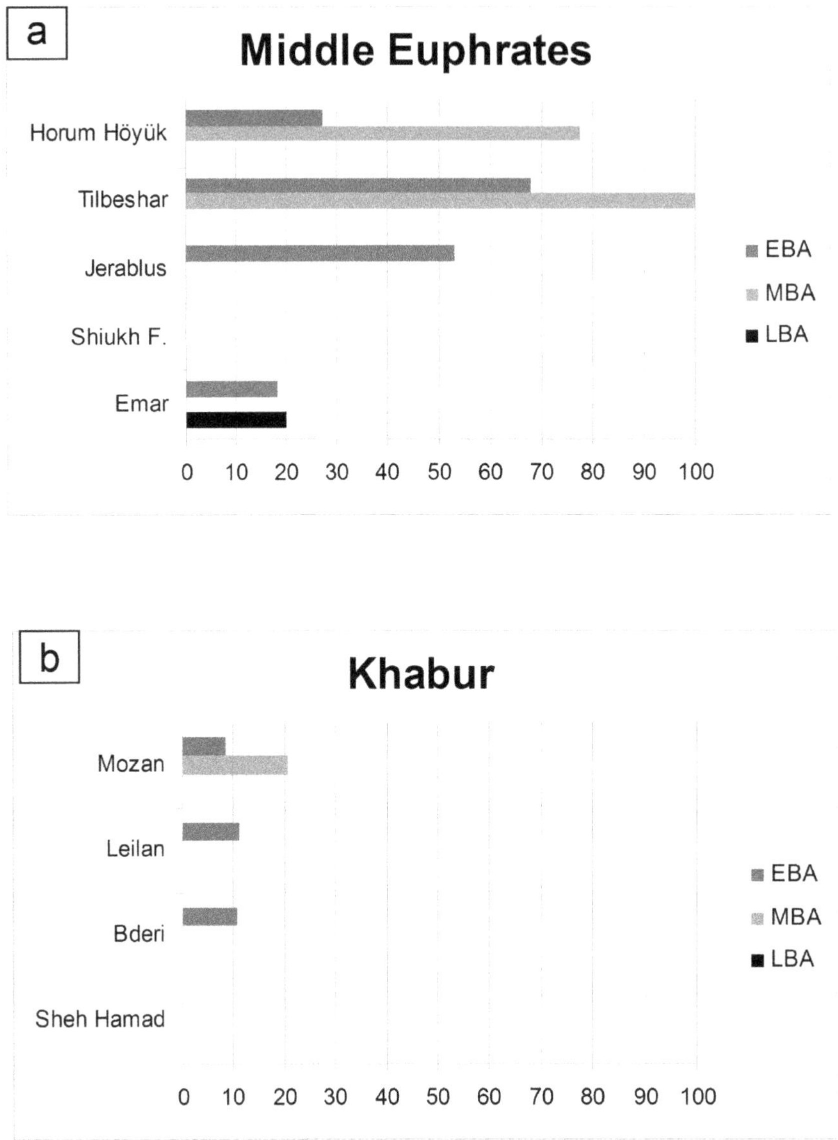

Fig. 6: Ubiquity of fruit trees in: a. the Middle Euphrates and b. the Khabur Basin.

5.2 Fruit tree cultivation

The percentages of fruit tree taxa indicate that, very early in the EBA of the investigated sites, fruit tree cultivation was hardly existent. It is only when the vegetation is paired with increased urbanisation that we observe a marked increase in cultivated taxa. This culminates in the MBA and then smaller proportions reappear in the LBA. In Horum Höyük, Tilbeshar, and Mozan, ubiquities show the same trend of increased presence of fruit trees in the samples from the EBA to the MBA (fig. 6). For Emar, however, no fruit tree charcoal fragments were found from the MBA strata and ubiquities in the EBA were equally high as in the LBA.

Overall, though, fruit tree cultivation seems to have been more important in sites located in areas with higher annual precipitation, in particular the Turkish Middle Euphrates sites. However, textual sources indicate that tree cultivation was more important than one would believe in the southern Middle Euphrates, like at Mari. Fig trees, pomegranates, poplars, junipers, and grapes were cultivated in specific areas at Mari, while olives, dates, and (sometimes) wine were imported products.[42]

[42] Michel 1996; Lafont 2000.

5.3 Riverine gallery forest

As mentioned before, there is low species diversity today within the riverine woodland of the Euphrates and Khabur River. Even more so, along the Upper Khabur Basin, riverine woodland is no longer present, as is also the case for the Wadi Jarrah. The anthracological data there indicated the presence of a gallery forest in the Bronze Age. Evidence has also been found for a riverine gallery forest in the Mid Holocene along the Wadi Jaghjagh.[43] Hence, these wadi environments must have looked very different in the past as compared to today. The archaeozoological data of the Khabur and Euphrates River support the presence of reed jungle, brushes, and trees along their benches because there have been finds of beaver, elephant, fallow deer, auroch, and boar as far south as Tell Seh Hamad and Emar.[44] The analysis of the diversity of riverine taxa within the samples indicates that, over time, at three of the five sites, there was a clear reduction in riverine diversity (fig. 7). In two northern Middle Euphrates sites, an opposite trend is visible. As for the Horum Höyuk site, this may be related to the increased use of the riverine vegetation and could, in fact, be an indicator of deforestation in the MBA. For Tilbeshar, it must be mentioned that the sampling could be unrepresentative because only four samples were investigated from the MBA. The southern Middle Euphrates site of Emar has a lower riverine taxa diversity than seen in northern Middle Euphrates sites; this may be due to the fact that the riverine forest was already degraded by the end of the EBA. There is, however, no indication of the increased impact of irrigation on the riverine vegetation through time. More precisely, there is no evidence for the consequences of reduced flooding on the vegetation. As is visible from Figure 8, in three of the four cases where diachronic data are available, *Tamarix* appears somewhat less frequently in the samples as compared to *Populus/Salix*; although the trend is not very strong. Even though the climate became drier, this could be on account of the May Euphrates discharge increasing, which then resulted in more extreme floods.[45] Nonetheless, overall there are relatively higher ubiquities of *Tamarix* towards the south, which may either be an indicator of local irrigation, or alternatively of its drier location. It is sometimes argued that the main Euphrates was too difficult to control and that because of this only smaller tributaries were used for irrigation. This was the case as described by texts for ancient Tutul (Tell Bi´a). But it is unclear whether all tells beyond the dry farming limit received water irrigation in this way.[46] A possible LBA irrigation canal has been found nearly twenty kilometres downstream of Emar.[47] Further south, adjacent to our study area, texts indicate that in the 18th century BC three canals irrigated the Mari kingdom, the remains of which have been discovered during regional surveys.[48] The canals can be traced back to the lower Khabur. For this region, Ergenzinger – Kühne[49] undertook extensive work to reconstruct and date the observed canal system. It probably is a somewhat later system than the one from Mari, with the canal on the eastern Khabur bench being in use from the 13th century BC onwards. However, this canal was probably fed by the Wadi Jaghjagh and not by the Khabur, and as such would not have had a major impact on the Khabur riverine vegetation.

5.4 Imported taxa

Local woody resources were accessible at all the sites during the Bronze Age, as is visible from the general absence of large proportions of imported wood. Although the MBA King, Yahdun-Lim, claims he was the first to cut cedars and pines in the coastal mountains, no evidence has been found for increased import of pines and cedars during that period. If *Pinus halepensis/brutia* was an import product (although we classified it in the "others" category), it was already traded from the EBA onward.

The findings of only low quantities of imported taxa, as revealed within this study, ultimately need to be differentiated. The samples detailed here are mainly from domestic contexts. Recent research at two MBA palaces has shown that distant woody resources were intensively used within palaces. For example, forty percent of the fragments identified from the Mozan Palace consist of imported *Pinus* wood,[50] while at Qatna (western Syria), the palace construction consisted mainly of *Cedrus* that must have been transported from relatively long distances (at least sixty kilometres).[51]

---

[43] Deckers – Riehl 2007a, Deckers, chapter 7 this volume.

[44] Gündem – Uerpmann 2003; Becker 2005.

[45] Riehl *et al.* 2008, fig. 1.

[46] Wilkinson 2004.

[47] Wilkinson 2004.

[48] Lafont 2000.

[49] 1991.

[50] Deckers 2010.

[51] Deckers, chapter 13 this volume.

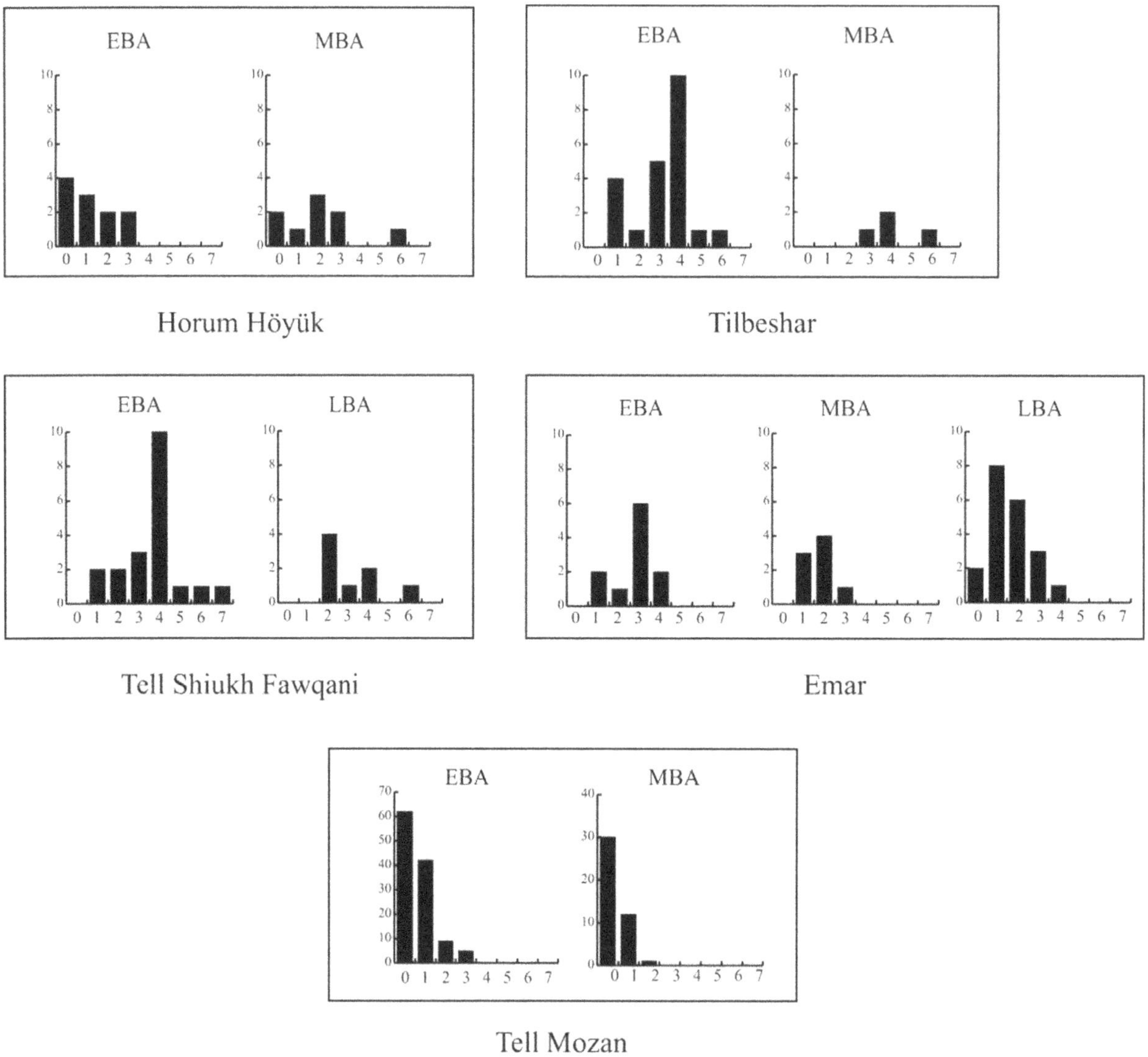

Fig. 7: Riverine species diversity within the samples. The x-axis shows the number of species within a sample, while the y-axis shows the number of samples.

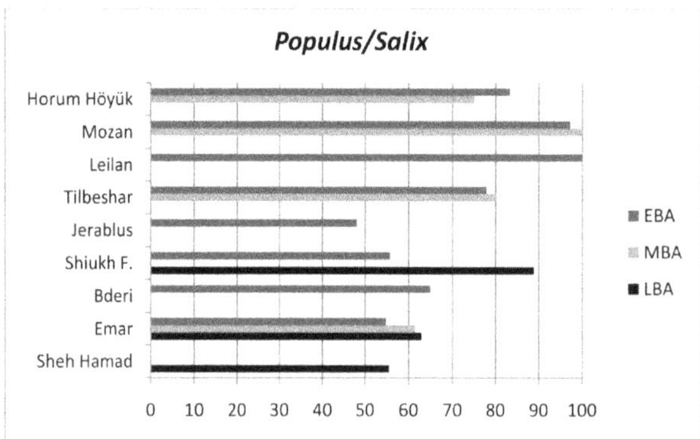

Fig. 8: *Populus/Salix* ubiquity percentages in proportion to *Tamarix*. Sites are listed from north to south (higher rainfall in the north).

6. Conclusion

The Bronze Age vegetation in the area under investigation was generally much richer than the one observed today. The riverine vegetation was a great deal more diverse, and oak park woodland could still be found in northern Syria, probably as far as Bderi on the Khabur. *Pistacia* (woodland) steppe occurred as far south as Emar. Hence, the present data suggests that massive degradation, including deforestation, occurred within our research area. There are some hints both for human and climatic impact on the vegetation through the Bronze Age but, still, by the end of this period, the region was much more covered in woody vegetation than is seen today.

Regarding human impact on the vegetation in the Bronze Age, in some cases, our trend analyses demonstrated a typical pattern for the exploitation of woody resources through time. Fields were probably preferentially located near the rivers, and the riverine forest was first cleared. After clearance of riverine vegetation took place, we can observe the next phase of increased exploitation as a consequence of the augmented demand for agricultural resources over time. This next phase is characterised by the increased use of land within the woodland steppe. Preliminary data on exploitation patterns of the oak park woodland in this area however, suggests sustainable wood use took place in the EBA and MBA.

We can especially observe this trend from clearance in the riverine forest followed by intensified use of park woodland at Tilbeshar from the EBA through the MBA, at Tell Shiukh Fawqani from the EBA until the LBA, and at Jerablus from early phases of the EBA until later phases of the EBA. In the case of Emar, the woodland steppe diminished in the LBA. This may have been caused by a combination of aridification and prolonged human impact, albeit discerning between the two remains difficult. Finally, the *Pistacia* to *Quercus* ratio indicates a shift northward for the *Pistacia* woodland steppe, due to more arid climatic conditions.

Bibliography

AKKERMANS, P.M.M.G. – SCHWARTZ, G.M.
2003 The Archaeology of Syria: from complex hunter-gatherers to early urban societies (c. 16,000 - 300 BC).

ALGAZE, G. – BREUNIGER, R. – KNUDSTAD, J.
1994 The Tigris-Euphrates Archaeological Reconnaissance Project: Final report of the Birecik and Carchemish Dam survey areas, Anatolica 20, 1-77.

BAR-MATHEWS, M. – KAUFMANN, A.
1998 Middle to Late Holocene (6,500 yr. period) paleoclimate in the eastern Mediterranean region from stable isotopic composition of speleothems from Soreq Cave, Israel, in: A.S. Issar – N. Brown (eds.), Water, environment and society in times of climatic change, 203-214.

BECKER, C.
2005 Small numbers, large potential – new prehistoric finds of elephant and beaver from the Khabur river/Syria, Munibe 57, 445-456.

BECKMAN, G.
1996 Texts from the vicinity of Emar in the collection of Jonathan Rosen; History of the Ancient Near East, Monographs II.

BOTTEMA, S.
1989 Notes on the prehistoric environment of the Syrian Djezireh, in: O.M.C. Haex – H.H. Curvers – P.M.M.G. Akkermans (eds.), To the Euphrates and beyond, archaeological studies in honour of Maurits N. van Loon, 1-16.

CHABAL, L.
1991 L'Homme et l'évolution de la végétation méditerranéenne, des âges des métaux à la période romaine, unpublished Ph.D. dissertation.

CHAVALAS, M.W.
1992 Ancient Syria: A historical sketch, in: M.W. Chavalas – J.L. Hayes (eds.), New horizons in the study of ancient Syria, Bibliotheca Mesopotamica 25, 1-21.

DECKERS, K.
2005 Anthracological research at the archaeological site of Emar on the Middle Euphrates, Syria, Paléorient 32.2, 152-166.

2010 Vegetation and wood use in the Bronze Age based on charcoals from Emar, in: U. Finkbeiner – F. Sakal (eds.), Emar after the closure of the Tabqa Dam, The Syrian–German Excavations 1996–2002, Vol. I: Late Roman and Medieval cemeteries and environmental studies, Subartu XXV, 225–244.

DECKERS, K. – RIEHL, S.
2007a Fluvial environmental contexts for archaeological sites in the Upper Khabur Basin (northeastern Syria), Quaternary Research 67.3, 337-348.
2007b An evaluation of botanical assemblages from the 3rd to 2nd Millennium BC in northern Syria, in: C. Kuzucuoğlu – C. Marro (eds.), Sociétés humaines et changement climatique à la fin du troisième millénaire: Une crise a-t-elle eu lieu en Haute Mésopotamie? Actes du colloque de Lyon, 5-8 décembre 2005, Varia Anatolica 19, 481-502.

DECKERS, K.
2010 Antracological research at the Early to Middle Bronze Age Settlement of Tell Mozan, in: P. Pfälzner (ed.) Ausgrabungen 1998–2001 in der Zentralen Oberstadt von Tall Mozan / Urkeš: The development of the environment, subsistence and settlement of the City of Urkeš and its region, SUN, Serie A, Vol. 3, 161–184.

ENGEL, T.
1993 Archaeobotanical analysis of timber and firewood used in the third millennium houses at Tall Bderi/northeast Syria, in: K.R. Veenhof (ed.), Houses and households, 105-113.

ERGENZINGER, P. – KÜHNE, H.
1991 Ein regionales Bewässerungssystem am Habur, in: H. Kühne (ed.), Die rezente Umwelt von Tell Sheikh Hamad und Daten zur Umweltrekonstruktion der assyrischen Stadt Dur-Katlimmu. Berichte aus der Ausgrabungen Tell Sheikh Hamad/Dur-Katlimmu 1, 163-190.

FRAYNE, D.R.
1997 Ur III Period (2112-2004 BC), Royal inscriptions of Mesopotamia 3/2.

FREY, W. – JAGIELA, C. – KÜRSCHNER, H.
1991 Holzkohlefunde in Dur-Katlimmu/Tall Seh Hamad und ihre Interpretation, in: H. Kühne (ed.), Die rezente Umwelt von Tall Seh Hamad und Daten zur Umweltrekonstruktion der assyrischen Stadt Dūr-Katlimmu, 137-161.

GALE, R. – CUTLER, D.
2000 Plants in Archaeology; Identification manual of vegetative plant materials used in Europe and the southern Mediterranean to c. 1500.

GEORGE, A.R.
2003 The epic of Gilgamesh.

GLENN, E.P. – NAGLER, P.L.
2005 Comparative ecophysiology of Tamarix ramosissima and native trees in western U.S. riparian zones, Journal of Arid Environments 61, 419-446.

GÜNDEM, C.X. – UERPMANN, H.-P.
2003 Erste Beobachtungen an den Tierknochenfunden aus Emar (Syrien) – Grabungen bis 2002, Baghdader Mitteilungen 34, 119-128.

HEMPEL, W.
2003 Letters to the King of Mari; A new translation with historical introduction, notes, and commentary.

LAFONT, B.
2000 Irrigation agriculture in Mari, in: R.M. Jas (ed.), Rainfall and agriculture in northern Mesopotamia. PIHANS 88, 129-145.

LEMCKE, G. – STURM, M.
1997 $\delta^{18}O$ and trace element measurement as proxy for the reconstruction of climate changes at Lake Van (Turkey): Preliminary results, in: H.N. Dalfes – G. Kukla – H. Weiss (eds.), Third Millennium BC climate change and Old World collapse, Global Environmental Change 49, 653-678.

MARGUERIE, D. – HUNOT, J.-Y.
2007 Charcoal analysis and dendrology: data from archaeological sites in north-western France, Journal of Archaeological Science 34, 1417-1433.

MICHEL, C.
1996 Le commerce dans les textes de Mari, in: J.-M. Durand (ed.), Amurru 1. Mari, Ebla et les Hourrites, dix ans de travaux. Actes du colloque international (Paris, Mai 1993), 385-426.

MILLER, N.F.
1985 Paleoethnobotanical evidence for deforestation in ancient Iran: a case study of urban Malyan, Journal of Ethnobiology 5.1, 1-19.

MOORE, A.M.T. – HILLMAN, G.C. – LEGGE, A.J.
2000 Village on the Euphrates.

PESSIN, H.
2004 Stratégies d´approvisionnement et utilisation du bois dans le Moyen Euphrate et la Damascène. Approche Antracologique comparative de sites historiques et préhistoriques, unpublished PhD dissertation.
2007 Analyses anthracologiques de deux sites du Moyen-Euphrate: Tilbeşar et Horum Höyük. Contribution à la problematique paléoclimatique de l´Holocène Moyen, in: C. Kuzucuoğlu – C. Marro (eds.), Sociétés humaines et changement climatique à la fin du troisième millénaire: Une crise a-t-elle eu lieu en Haute Mésopotamie? Actes du colloque de Lyon, 5-8 décembre 2005, Varia Anatolica 19, 557-572.

POWELL, M.A.
1992 Timber production in presargonic Lagaš, Bulletin on Sumerian Agriculture 6, 99-122.

RIEHL, S. – BRYSON, R. – PUSTOVOYTOV, K.
2008 Changing growing conditions for crops during the Near Eastern Bronze Age (3000-1200 BC): The stable carbon isotope evidence, Journal of Archaeological Science 35.4, 1011-1022.

RIEHL, S.
2010 Maintenance of agricultural stability in a changing environment — the archaeobotanical evidence at Emar, in: U. Finkbeiner – F. Sakal (eds.), Emar after the closure of the Tabqa Dam, The Syrian–German Excavations 1996–2002, Vol. I: Late Roman and Medieval Cemeteries and Environmental Studies, Subartu XXV, 177–224.

RISVET, L.
2005 Settlement, economy, and society in the Tell Leilan region, Syria, 3000-1000 BC., unpublished PhD dissertation.

SMART, T.L. – HOFFMAN, E.S.
1988 Environmental interpretation of archaeological charcoal, in: C.A. Hastorf – V.S. Popper (eds.), Current Palaeoethnobotany, 165-205.

TARAQJI, A.F.
1997 De relatie tussen Boven-Mesopotamië (de Jezireh) en de andere delen van Syrië in de Midden-Brons periode, in: P. Talon – K. Van Lerberghe (eds.), In Syrië. Naar de oorsprong van het schrift, 135-137.

THALEN, D.C.P.
1979 Ecology and utilization of desert shrub rangelands in Iraq.

WESTENHOLZ, J.G.
2000 Cuneiform inscriptions in the collection of the Bible Lands Museum Jerusalem; The Emar Tablets.

WICK, L. – LEMCKE, G. – STURM, M.
2003 Evidence of Late Glacial and Holocene climatic change and human impact in eastern Anatolia: High-resolution pollen, charcoal, isotopic and geochemical records from the laminated sediments of Lake Van, Turkey, The Holocene 13.5, 665-675.

WILKINSON, T.J.
2004 Excavations at Tell Es-Sweyhat, Syria, Vol. 1., On the margins of the Euphrates. settlement and land use at Tell es-Sweyhat and in the Upper Lake Assad Area, Syria, Oriental Institute Publications 124.

WILLCOX, G.
2002 Evidence for ancient forest cover and deforestation from charcoal analysis of ten archaeological sites on the Euphrates, in: S. Thiébault (ed.), Charcoal analysis; Methodological approaches, palaeoecological results and wood use, BAR Int. Series 1063, 141-145.

2010 Preliminary results from Tilbeshar (in condensed form) from the 2000 season, in: G. Willcox (ed.), George Willcox (Archaeobotanist). Retrieved 10/06/2010 from http://pagesperso-orange.fr/g.willcox/first.htm.

ZOHARY, M.
1973 Geobotanical foundations of the Middle East, Vol. 2.

ZOHARY, D. – HOPF, M.
2000 Domestication of plants in the Old World.

4. Geoarchaeology in the Tell Mozan surroundings

Katleen Deckers and Konstantin Pustovoytov

1. Introduction

Tell Mozan is a well known settlement mound with an associated lower town located in the Upper Khabur Basin of northeastern Syria, 6.2 km southeast of Amouda (fig. 1). Its beginnings are as yet unknown but the site dates back to at least the early part of the fourth millennium BC.[1] A major occupation phase dates back to the second half of the third millennium BC. More specifically, Mozan has been identified as the important late third millennium Hurrian capital of Urkiš. Unlike many other sites in this area, Mozan was continually occupied at the end of the third millennium BC. It was probably an ally or vassal of the Akkadian Empire around 2300 BC. Steady occupation at the site between 2200 and 1900 BC may have related to the then newly established kingdoms of Urkish and Nawar. During the second millennium BC the city passed to the rulers of Mari and became a vassal.[2] The city was abandoned about halfway through the second millennium BC. Archaeological surveys indicate that the main occupation phase of the lower town was in the Early Jezira III period (ca. 2700-2250 BC); and it is thought that the outer wall was constructed around the lower town in this period. In the early Middle Bronze Age (from ca. 2100 until 1600 BC) the lower town was less densely inhabited and it has been suggested that the lower town wall was no longer existent.[3]

The site of Tell Mozan lies within the dry farming zone, which has a mean annual rainfall of approximately 420 mm/year. Today, hardly any trees grow within the area, which is intensively used as agricultural land for cereal cultivation. The dominating presence of oak, the presence of acorns, and the charcoal from small oak branches in the site's charcoal samples suggest that the oak-Rosaceae park woodland had a more southward distribution in the mid third millennium BC than presently.[4]

2. Aims of this study

Investigations of satellite images from the sixties indicate that the landscape immediate to Tell Mozan has changed drastically and several features surrounding the site need further geoarchaeological investigation to improve comprehension of their function. The aims of this study consist of understanding Tell Mozan in its fluvial landscape setting and resolving the function of the ditch surrounding the site.

In the field today, it is nearly impossible to discern how Mozan would have been provided with water in the past. Interestingly, on CORONA satellite image D025-055 1105-1 FWD, from the 5th of November 1968, a wadi is seen. It can be traced from the northeast to the southwest, through the western lower town of Mozan and passing the present day village of Mozan to the east (fig. 1). Just north of the lower city wall, another branch of the same wadi seems to follow the contour of the depression around the lower city wall, passing the modern village of Mozan at its western side. Both channels join again at about 350 m southwest of the modern village of Mozan, at the lower town's outer edge. The depression surrounding the lower town is a feature that is visible on satellite images because there the soil is moister. But it is also still visible in the field, especially to the south of Mozan where it is nearly a 100 m wide depression.

The exact kind of stream that was passing through Mozan needs to be investigated, as well as whether there existed a connection between this stream and the ditch surrounding the site at the time of the occupation. Additionally, it is also important to gain insight into the broader evolution of the landscape, in order to improve understanding of the palaeoenvironmental conditions in the sustaining area

1 Buccellati – Kelly-Buccellati 1999.

2 Buccellati – Kelly-Buccellati 1999.

3 Thompson-Miragliuolo 1988; Pfälzner – Wissing 2004.

4 Deckers 2005; Deckers – Riehl 2007; see also Deckers – Pessin, chapter 3 this volume.

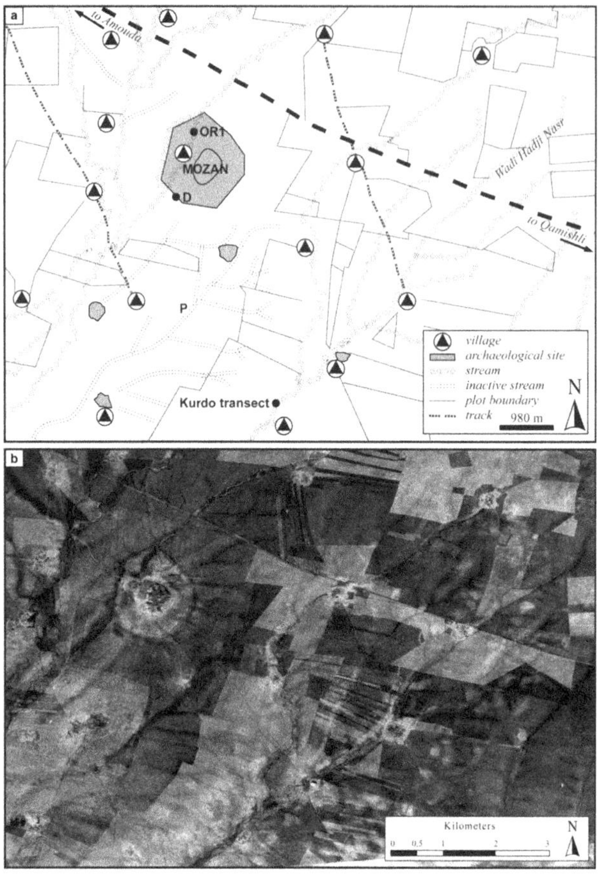

Fig. 1a: CORONA satellite image of the tell Mozan surroundings; 1b. Map based on the CORONA image with all investigated features indicated.

of Tell Mozan. Of particular interest is the straight shape of a drainage feature some 1 km southeast of the lower town wall at Mozan, visible as darker lines (P on fig. 1a). It is questionable as to whether it could have been related to artificial irrigation.

3. Methods

Fieldwork took place in the frame of the Tell Mozan archaeological excavations under the directions of Prof. Giorgio Buccellati, Prof. Marilyn Kelly-Buccellati, and Prof. Peter Pfälzner. A wadi profile (OR1) was investigated in 2006, while a section from the ditch (from this point on referred to as 'profile D') was investigated in 1999. Both profiles were dug by hand to a depth of more than 3 m from the present surface. Additionally, the profiles from the Wadi Hadji Nasr (also referred to as 'Kurdo transect') were retrieved in 2003 by a hand-auger that reached a depth of approximately 4-5 m. Another of the profiles was studied in the quarry, known as the Kavvla locality, where many gravels were exposed. All profiles were described in the field for their sedimentological and soil formation characteristics.

Special care was taken to retrieve samples that would help us in gaining chronological insight. For example, pottery sherds were collected and typologically investigated. Moreover, optically stimulated luminescence (OSL) sediment samples and radiocarbon samples consisting of pedogenic carbonate, charred organic material, and soil organic matter were also retrieved. The OSL samples from profile OR1 were dated by Dr. Alexandra Hilgers[5] and preliminary dating work was undertaken by Dr. Markus Fuchs on the sample from the Kurdo profile. Finally, samples for radiocarbon dating were analysed in Kiel by the Leibniz Labor für Altersbestimmung und Isotopenforschung, University of Kiel (Table 1).

Micromorphological and bulk sediment samples were collected to help in furthering knowledge of the immediate depositional and soil forming processes. Samples for micromorphological study were collected

[5] see Hilgers, chapter 11 this volume.

Table 1: AMS radiocarbon dates for profile D with isotopic values and calibration. Calibration with Calpal online 2009.

Sample number	Material	Depth below soil surface	Details of dated material	Radiocarbon age (BP)	Calibrated date (BC)	$\delta^{13}C$	Comment
KIA30311	sediment matrix	180-197 cm	a. insoluble organic-rich fraction	4730 + 170/-160	3446 ± 222	-25.32	possibly too low C content
			b. charcoal	4280 ± 70	2887 ± 113	-22.82	-
			c. humic acid from charcoal	4045 ± 40	2580 ± 68	-25.11	possible influx of younger C
KIA30312	charcoal	180-197 cm	-	4235 ± 55	2807 ± 89	-22.47	-
KIA30313	non-floating organic particles	180-197 cm	-	4500 + 90/-80	3191 ± 140	-23.07	possibly too low C content
KIA30314	pedogenic carbonate	210-230 cm	-	38470 +1380/-1180	N.A.	-8.55	-
KIA30315	pedogenic carbonate	155-165 cm	-	2350 ± 35	442 ± 44	-8.76	-
KIA28162	pedogenic carbonate	80-100 cm	oldest lamina from coatings on shard	2625 ± 25	806 ± 8	-10.55	-
KIA29318	pedogenic carbonate	80-100 cm	oldest lamina from coatings on shard	2815 ± 45	979 ± 55	-14.54	-

as undisturbed blocks from profile D. They were impregnated and processed into thin sections by Thomas Beckman and were analysed by Prof. Paul Goldberg.[6]

Sedimentological analysis was undertaken at the University of Tübingen for the bulk samples from section D. Calcium carbonate content, pH, and particle size analysis were measured according to the procedures described in Beck *et al.*[7] The C, N, and S percentages were measured using the Vario E1III elemental analyser.

The straight shape of a drainage feature was studied through exaggerated 3D-SRTM data that was overlain by the CORONA satellite images. This was done because the topography in the field is very flat.

4. Results

4.1 Profile OR1 (figs. 1, 2, and 3)

At the place where a palaeostream was visible on satellite images, a slight depression can be observed in the field. Here, only at one locale are the wadi bed and benches still largely present.

Pit OR1, at 2×2 m and with a depth of roughly 3.3 m, was made just east of the filled-in depression of the former wadi and is within the border of the lower town of Mozan (fig. 1). The upper 1.7 m consists of prismatic silty clay sediments in which relatively strong soil formation took place (fig. 2). This is mainly evidenced by strong calcium carbonate accumulation, in the form of carbonate nodules between depths of 1 and 1.7 m, accumulation of organic matter at the land surface and a system of vertical cracks within the uppermost 30 cm. Oxidation stains are observed extending between 1 and 2.25 m deep. The lower part of the profile contains two layers of poorly sorted gravels (figs. 2, 3b, and 3c). The lowermost gravel is in a clayey sand matrix, is imbricated, and contains no artefacts. The clasts are rounded to subrounded and up to 6 cm in size. The lowermost gravel is somewhat better sorted and oriented (albeit limited in orientation) than the uppermost gravel. The latter contains many archaeological artefacts, is dark brown and is very rich in organic matter. The clasts of the uppermost gravel are less rounded, being subrounded to subangular. Both gravels were followed by clayey silt deposits that contain some clasts (figs. 2 and 3b).

Pottery sherds were collected in artificial layers. Most of them are dated to the third millennium BC, although there were also some fourth millennium BC sherds ('local ware,' no fine ware).[8] Two OSL dates were analysed for the lowermost layers; the first has a date between 6 and 7.3 kyr BP (C-L2013) and the other is of Pleistocene age (C-L2014).[9]

4.2 Profile D (figs. 1, 4, and 5)

The upper horizon of this profile consists of an 18 cm thick plough horizon. From a depth of 18 until 70 cm a combination of Ah and AB horizons were found. Isolated crevices from AB extend down to about 1 m in depth.

Between a depth of 70 and 183 cm, there is a greyish-brown zone referred to as a Bk horizon. Within this depth interval there are krotovinas from several phases. Although the material of this layer looks lithologically uniform, consisting of silty clay, it was most likely deposited in several phases as some horizontal cracks (e.g. at depths of 115 and 143 cm), horizontally oriented sherds, and a subhorizontal krotovina suggest. Of special note is the find of dung in the micromorphological slides from the lowermost part of this section.[10] The upper 183 cm of the profile must have been deposited sometime before 1000 BC; this is the age provided by the oldest laminae of the carbonate coatings on the pottery sherd found at a depth of 1 m.

Between 183 and 197 cm a possible Ahb horizon was found (i.e. a buried Ah). It differed from the material of the overlying Bk horizon in having a stronger grey-color and an increased amount of ceramic sherds, fine charcoals, and aggregates of relatively dense charred matter. It is also marked by an S-peak just like that of the present day surface. The micromorphological samples indicate biological reworking.[11] Secondary carbonate accumulation also took place in the entire Ahb horizon. Additionally, the Ahb horizon has a constant thickness that remains unchanged, even when obvious and slight microrelief irregularities on the palaeosurface were observed, and the pottery sherds did not have a special orientation. This may have

[6] see Goldberg, chapter 5 this volume.

[7] Beck *et al.* 1995.

[8] investigated by Prof. M. Kelly-Buccellati, pers. comm. 2006.

[9] see Hilgers, chapter 11 this volume.

[10] see Goldberg, chapter 5 this volume.

[11] see also Goldberg, chapter 5 this volume.

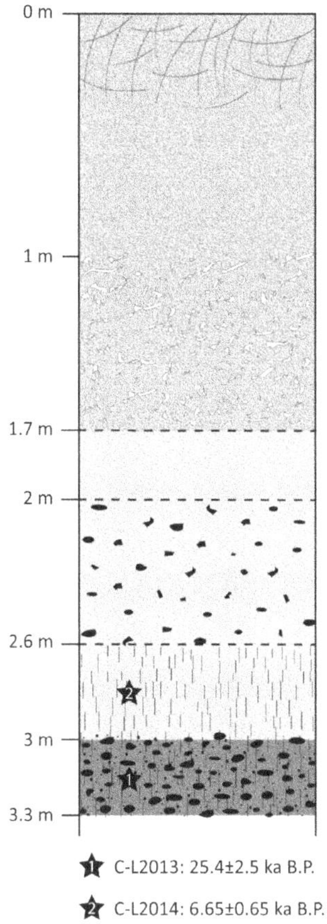

Fig. 2: Schematic representation of the OR1 profile.

Fig. 3a: Photograph taken from Tell Mozan to the north. Profile OR1 was investigated where tent is visible (indicated by arrow); 3b. Image of OR1; 3c. Image of lower part of profile OR1 with two gravel layers.

Fig. 4a: Photograph of the depression with arrow indicating the location of profile D; 4b. Field image of the profile. It has a depth of ca. 3 m.

been due to systematic turbation of the palaeosurface to a certain depth (ca. 15 cm), as can be the case with ploughing. Several dates were obtained for this horizon, ranging between 3446 ± 222 BC (KIA30311a) and 2580 ± 68 cal. BC (KIA30311c). However, some dates are problematic (see also comments in Table 1): More specifically, the dates may be too young for the humic acid, due to leaching, and for the uncarbonised organic remains of which samples were extremely small, dates may be too old. Interestingly, two

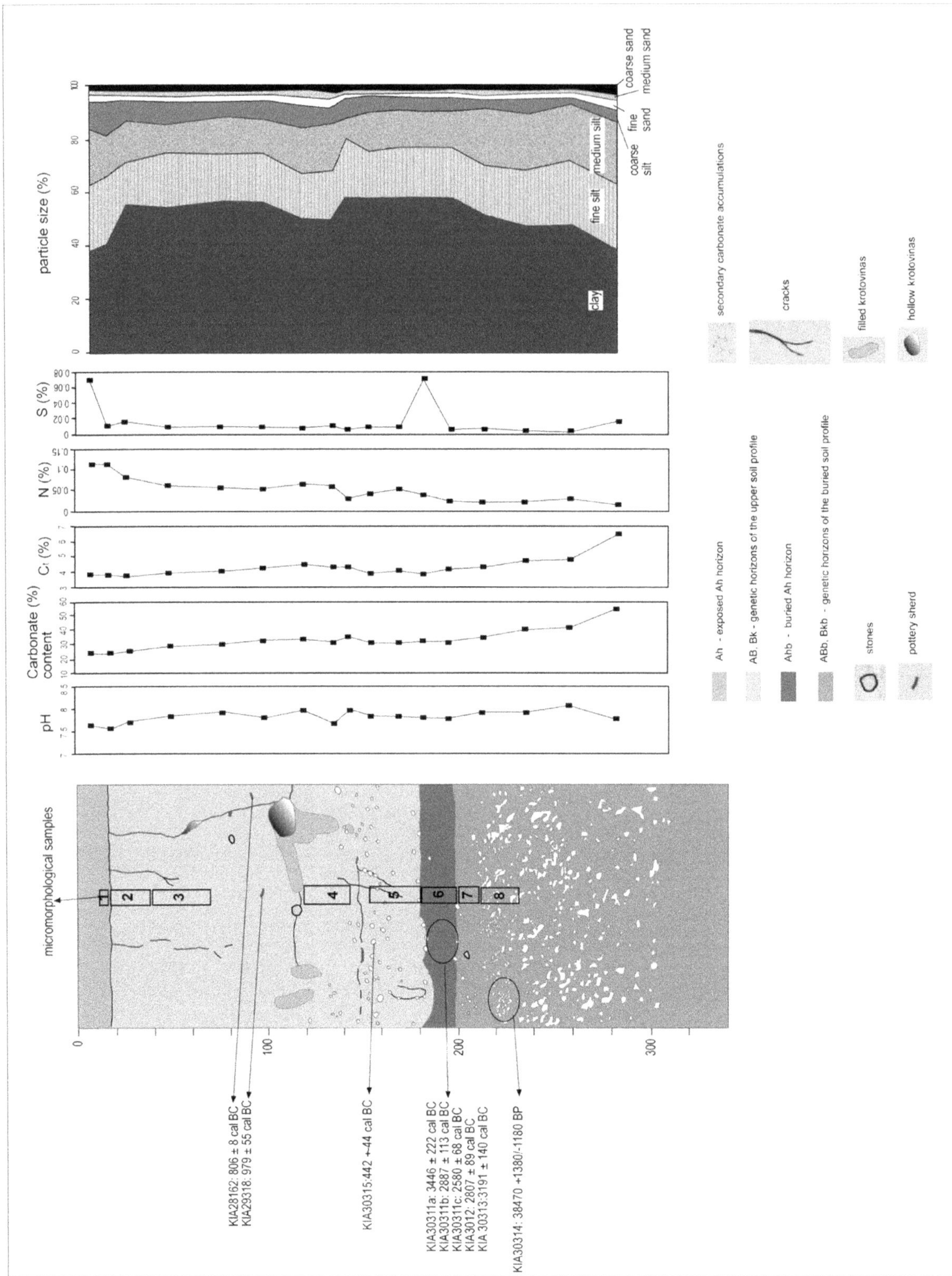

Fig. 5: Schematic representation of profile D with physico-chemical properties.

charcoal insoluble fraction samples for this horizon provide an approximate date of 2800 calibrated BC and suggest that it may have been exposed at the surface about this time. Finally, the oldest dated pedogenic carbonate coating from the soil above indicates that this level was covered, at the latest, by about 1000 BC (KIA29318).

The deepest horizons, BC (197-210 cm) and BCk (210-330 cm), differ in color from all the overlying horizons with their more intense reddish component. The two horizons are very similar in their properties but BCk incorporates numerous carbonate nodules. The BC horizon is marked by slaking crusts.[12] Below 210 cm there are no artefacts or charcoals and a Pleistocene date was obtained here through several carbonate nodules.

4.3 The Wadi Hadji Nasr transect (figs. 1, 6, 7, and 8)

Within the quarry site of Kavvla, which is located within a river terrace, a large area of gravel was exposed. One profile, Kurdo (fig. 7), was studied in detail because an archaeological site was incised into its gravel deposits. The examined profile consists of poorly sorted, rounded gravel in a sandy matrix. At its topmost section, it is composed of silty sand, while lower down it is of rather coarse sand. There is some grading visible at a depth of about 1.5 m where the gravel is fining upwards. Laterally, the gravel gives way to laminated sands that are interrupted by gravel layers.

An OSL sample was retrieved from the gravel at a depth of 1 m. Since the profile contains strong evidence for $CaCO_3$ accumulation, especially visible as thick $CaCO_3$ crusts on the lower side of clasts and with the presence of calcrete between 1 and 1.85 m, it is assumed these sediments were deposited during the Pleistocene. The OSL date acquired probably puts this section within the Late Pleistocene.[13]

Within the present valley of the Wadi Hadji Nasr, four boreholes were investigated (fig. 6), each located some 15 m from another. Borehole 4 is located nearest to the Hadji Nasr, within a distance of 45 m.

Kurdo borehole 1 has a depth of 3.9 m (fig. 8). The upper 3.4 m is comprised of fine-grained sediments that occasionally contain clasts in the size order of between 1 and 4 cm. They are angular to subangular. From a depth of 3.4 m onwards gravel was reached. Between a depth of 1.1 and 3.9 m, $CaCO_3$ accumulation and manganese flecks occurred. Furthermore, a sherd was found at a depth of roughly 0.75 m and a charcoal fleck occurred at a depth of around 1.5 m.

Kurdo borehole 2 (fig. 8) consists of 5 m of clayey silt deposits that sometimes include angular chalks of between 1 and 3 cm in size. From a depth of about 2 m, $CaCO_3$ accumulation is visible in the shape of $CaCO_3$ nodules; manganese flecks are also present from here onwards.

Kurdo borehole 3 (fig. 8) has a depth of 4.1 m and is composed of silty clay until a depth of 3 meters, with occasional angular clasts present. From 3 m onwards the sediment is clayey sand. There is a pronounced $CaCO_3$ accumulation between a depth of 0.8 and 3 m. Lastly, one charcoal fleck was found at a depth of about 1.9 m.

The final borehole, Kurdo borehole 4 (fig. 8), has a depth of 5 m and consists entirely of clayey silt deposits with occasional small angular clasts. In the lower 1.7 m, clasts are somewhat larger with sizes up to 3 cm. Finally, between depths of 1.2 and 2.7 m $CaCO_3$ accumulation and manganese flecks occur.

4.4 The regular drainage feature

Due to the fact that the landscape in the immediate Mozan surroundings appears relatively flat, exaggerated 3D-SRTM data provide valuable information (fig. 9). Interestingly, the drainage features do not seem to have been irrigation channels since we would expect the water to be distributed from the larger channel to the smaller channels and not the other way around, as is the case here. For this reason, these features seem rather to be a natural palaeo-drainage pattern. A question that also arises at this point is whether they might have been connected with the Wadi Hadji Nasr. The topography as seen from the exaggerated 3D-SRTM data does not indicate a clear connection point and there seems to be a topographic boundary between both drainage systems. If there once was a connection between both streams, it would have been

[12] see Goldberg, chapter 5 this volume.

[13] Dr. M. Fuchs investigated two OSL samples from the geoarchaeological survey. One derived from this profile, Kurdo 1, and the other from the Hamidi sequence, which was studied later in greater detail for its chronology (see also chapter 10 this volume). Unfortunately, it appears that the OSL sample from Kurdo 1 was most likely swapped with the Hamidi sequence sample. This is speculated for the following reasons: the intensive $CaCO_3$ accumulation within the Kurdo 1 profile and the abnormally high-flow velocity indicates a Pleistocene age. For this reason, a date of 5.75 ± 1.94 kyr BP is very unlikely. Furthermore, this date fits very well with all other dates from the Hamidi sequence.

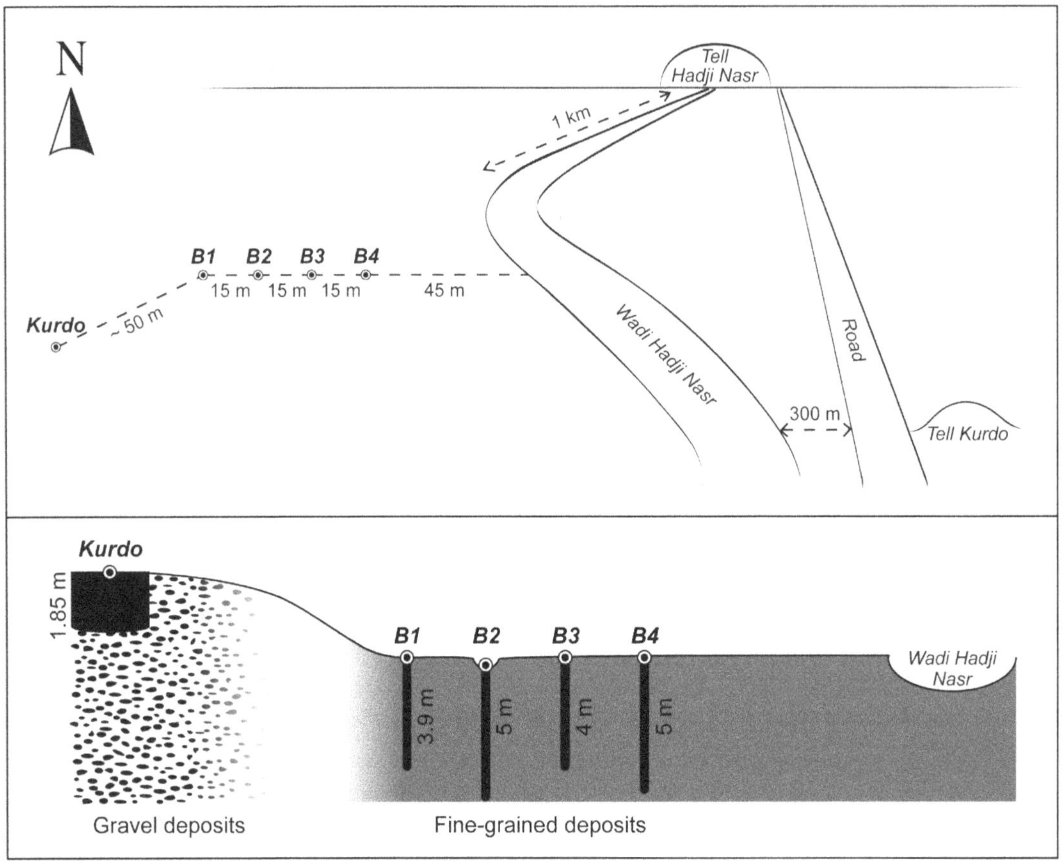

Fig. 6: Schematic representation of the transect from Wadi Hadji Nasr.

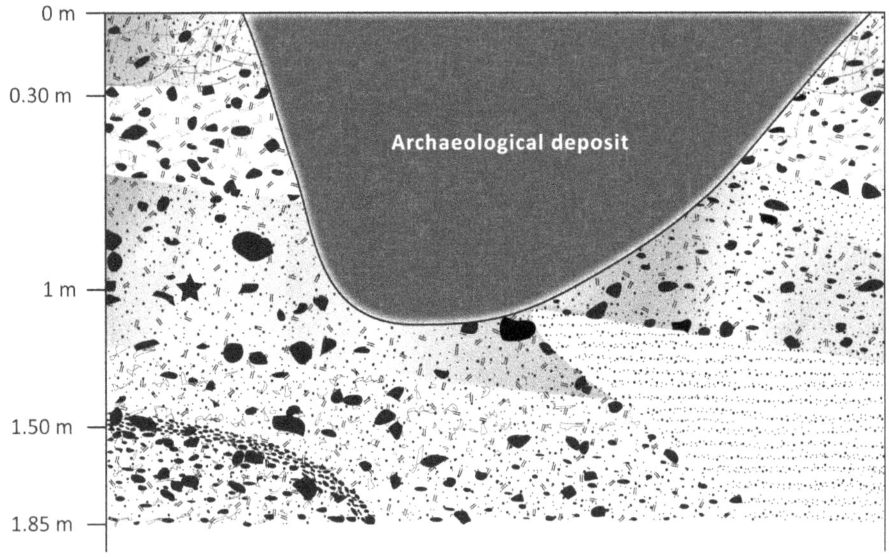

Fig. 7: Schematic representation of the Kurdo profile.

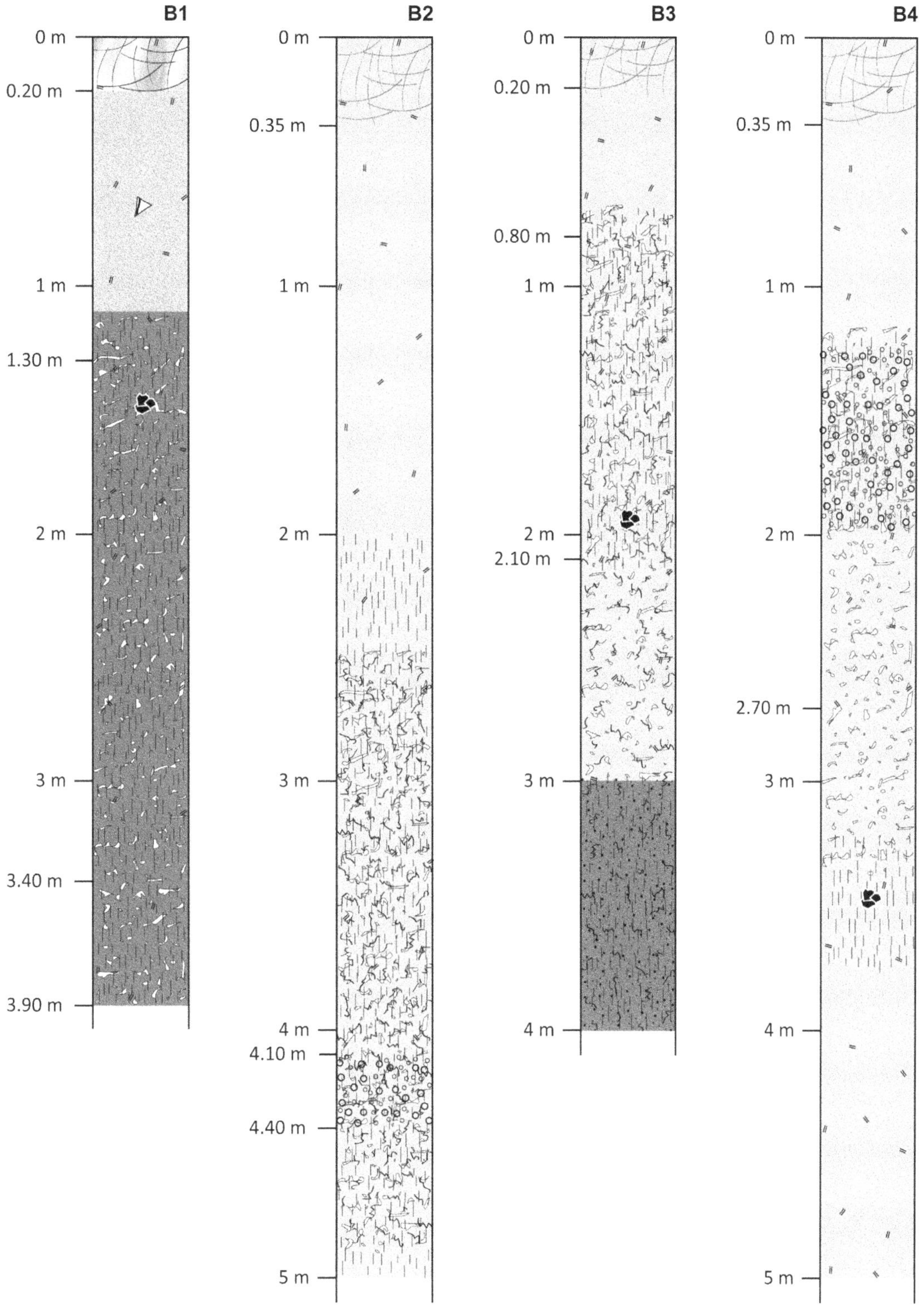

Fig. 8: Detailed schematic representation of the four boreholes along the transect from Wadi Hadji Nasr.

Fig. 9: 3D-SRTM image indicating the drainage system southeast of Mozan was not an irrigation system.

before the incision of the landscape took place and/or tectonic impact would have needed to occur in order to justify the topographic boundary.

5. Interpretation of the landscape's evolution

At OR1 we find poorly sorted, non-oriented, non-graded gravels, which indicate that they were deposited in flashflood-like conditions. In particular, the uppermost gravel indicates that the sediments were not transported over a long distance. The profile contains many artefacts because it is located within the lower town of Mozan, which was intensively inhabited, albeit only from 2700-2250 BC. Therefore, the upper artefact-rich 2.8 m must have been deposited sometime during or after the third millennium BC, while the gravel without sherds must have been deposited before the foundation of the lower town. The OSL dates do not contradict this. Date C-L2014 matches especially well, marking the deposition at 6.65 ± 0.65 kyr BP. The uppermost 1.7 m of silty clay sediments seems to have been deposited some time ago, allowing for deep soil formation. Additionally, carbonate accumulations are found here between 1 and 1.7 m and iron oxidation stains exist from 1 to 2.25 m, the latter indicate a temporary wetting and drying of the soil. According to the older men of the village of Mozan, some 25 years ago fields in this area could not be cultivated because they were flooded heavily during the winter and the soil was too moist.

The formation of the pediment visible in profile D can be reconstructed through recognising a series of processes that most likely took place. Firstly, the Pleistocene soil was incised, most probably as a result of brick quarrying activities. As a result of this process, a depression came into existence that probably temporarily contained water. Subsequently, portions of organic-rich anthropogenic material accumulated at the bottom of the depression. This level was at the surface for a while and was probably ploughed. The ^{14}C dates suggest this was around 2800 BC. Finally, this level was buried, at the latest, sometime before 1000 BC. Several generations of alluvial sediments, including a lot of anthropogenic material, filled in the depression. They may represent some kind of "plough wash." The timing of these events is sometime between 2800 BC and 1000 BC. During this period, the depression as well as its surroundings may have been continuously cultivated. The coproliths within the lowermost reach of this level may indicate the use of manure for cultivation. An alternative explanation for the sediments at this level is that they may represent dumped fluvial sediments that were redeposited or garbage from the settlement that was bioturbated and in which incipient soil formation could take place. The radiocarbon dates available do not allow making an ultimate conclusion about the age of the uppermost 0.5 m of the profile. It cannot be older than the ceramics at 80-100 depth (ca. 2700 BC) but its youngest limit remains uncertain.

6. Discussion

6.1 Alluvial landscape context at Mozan

The wadi that passed through the lower town of Mozan seems to have been an intermittent stream that existed before the occupation of the lower town (there is sedimentological evidence for this sometime around the fifth millennium BC) and probably also during, or perhaps after, the occupation of the lower town in the third millennium BC.

According to the local people, the reduction in water availability within the wadi might be related to damming in Turkey but no evidence for this can be found on recent satellite images. In fact, it is almost impossible to trace back the location of the wadi over the whole stretch of land in these images, while this is still possible on satellite images from the sixties. It also is of note that neither the satellite images from the sixties nor the Russian topographic maps indicate that this wadi had its source in Turkey; rather, they show it originating in the upper north of Syria. It is therefore likely that the cause for its disappearance relates to the lowering of the groundwater level, which earlier must have come into contact with the wadi bed.

Since this stream was intermittent it is logical that the ditch surrounding the lower town of Mozan could not have been perennially filled with this water. Moreover, the ditch profile does not indicate that it contained stagnant water. For this reason, the ditch's potential role in offering strategic protection can also most likely be excluded. What is likely is that the ditch was used as a moister agricultural area. However, it is of interest that sometimes the ditch was flooded during rainfall events and that it contains a mixture of alluvial and colluvial deposits.

The drainage feature southeast of Mozan is not an artificial irrigation feature, rather it seems to be a (palaeo-) drainage. Furthermore, no obvious connection with the Wadi Hadji Nasr was found. The Wadi Hadji Nasr must have been a considerable stream in the Pleistocene as its gravels, indicating high velocities, suggest. Following the Pleistocene, sometime within the Holocene, a phase of strong incision took place and more than 5 m of fine-grained material filled in the valley. This hints at greater topographic differences in the past than can be seen today.

6.2 Erosion in the Tell Mozan surroundings

Rapid sediment accumulation took place at the location of profile D sometime between 2800 and 1000 BC, which might have been a result of intensive land use around the Bronze Age town. The growth of urban centres put pressure on its hinterland to cover its subsistence needs. Resource calculations indicate that within a radius of roughly 5 km around the town of Mozan, agricultural land must have been under use (if we assume normal population densities).[14] While the charcoal data from Mozan suggests a more southerly distribution of deciduous oak during the Early Bronze Age, the charcoal-to-seed ratios from Mozan point to the use of dung as fuel and therefore could be taken to indicate a reduced availability of woody resources for this period.[15] The charcoal data from Mozan suggest an intensification of human impact on the vegetation at the start of the Middle Bronze Age, which may somehow be linked to the development of the Hurrian centre located at Mozan during this time.[16] It is therefore possible that a large part of the sediments were accumulated in the early Middle Bronze Age.

Additionally, the Wadi Hadji Nasr Holocene fill indicates that considerable erosion took place during the Holocene. Unfortunately, as of yet, these sediments could not be numerically dated. What is clear, nevertheless, is that a large part of the erosion took place relatively locally. This is made evident through the existence of a colluvial proportion within the fluvial fine-grained deposits, as seen in the occasional presence of clasts that are frequently angular.

6.3 Agricultural practices in the Tell Mozan surroundings

There are indications for ploughing and fertilising (using manure) from the ditch profile. Although it is known that the plough was in use at this time, geoarchaeological indications for its use have up to now not been found. For example, contemporary texts from Tell Beydar, some 50 km southwest of Mozan, mention the amount of plough animals used in Beydar, as well as the number of draft animals allocated to six smaller settlements around the site.[17]

Based on the intensive off-site shard scatters surrounding third millennium BC sites in northern Mesopotamia, Wilkinson has suggested that often burned dung was used as manure. Furthermore, because it was used as fuel, it had to pass through the settlement and contained many other types of settlement debris.[18] This could explain the large amounts of anthropogenic inclusions in the sediment of profile D.

7. Conclusion

Geoarchaeological research conducted in the Tell Mozan area has proven to elucidate several burning questions. To begin with, it has made it clear that Tell Mozan was located along an intermittent stream, even though today little is visible of this stream. Secondly, the inactive, regular drainage pattern southeast of Mozan was not an irrigation system, but represents a natural palaeo-stream with a regular pattern. The ditch around the Bronze Age settlement also did not permanently contain water. It was more likely used as an agricultural field where the soil was moister. Furthermore, the uppermost 2 m of the profile represents a record of

---

[14] Deckers – Riehl 2008, Deckers – Drechsler, chapter 2 this volume.

[15] Deckers in press, see also Deckers, chapter 12 this volume.

[16] Deckers – Riehl 2007.

[17] Ismail *et al.* 1996, 3; see also Widell 2005.

[18] Wilkinson 1990; Wilkinson – Tucker 1995.

accumulation that includes anthropogenically-derived, redeposited, and bioturbated debris. This accumulation started at around 2800 calibrated BC or prior to that and was especially intensive sometime between 2800 and 1000 BC. It may have been a sedimentological response to agricultural exploitation of the site's surroundings; for example, during the urban explosion of the Early Bronze Age or in the early Middle Bronze Age when increased human impact on the woody vegetation took place. Finally, it has been established that along the Wadi Hadji Nasr intensive fluvial deposition took place during the Holocene, with a local fraction of colluvial sediment.

Bibliography

BECK, R. – BURGER, D. – PFEFFER, K.-H.
1995 Laborskript. Ein Handbuch für die Benutzer der Laboratorien der Physischen Geographie der Universität Tübingen.

BUCCELLATI, G. – KELLY-BUCCELLATI, M.
1999 Das archäologische Projekt Tall Mozan/Urkeš, Mitteilungen der Deutschen Orient-Gesellschaft zu Berlin 131, 7-16.

DECKERS, K.
2005 Anthracological research at the archaeological site of Emar on the Middle Euphrates, Syria, Paléorient 32.2, 152-166.
2011 Bronze Age archaeological sites in the landscape: On the former distribution and density of deciduous oak in northern Syria, in: Conard, N.J.; Drechsler, P. and Morales, A. (eds) Between Sand and Sea. The Archaeology and Human Ecology of Southwestern Asia, 177-190.

DECKERS, K. – RIEHL, S.
2007 An evaluation of botanical assemblages from the 3rd to 2nd millennium BC in northeastern Syria, Varia Anatolica XIX, 481-502.
2008 Tells of the Upper Khabur: their hydrological and agrarian contexts, Paléorient 34.2, 173-189.

ISMAIL, F. – SALLABERGER, W. – TALON, P. – VAN LERBERGHE, K.
1996 Administrative documents from Tell Beydar, seasons 1993-1995. Subartu II.

PFÄLZNER, P. – WISSING, A.
2004 Urbanismus in der Unterstadt von Urkes, Mitteilungen der Deutschen Orient-Gesellschaft zu Berlin 136, 41-86.

THOMPSON-MIRAGLIUOLU, J.
1988 The outer city: Introduction and surface collection, in: G. Buccellati – M. Kelly-Buccellati (eds.), Mozan 1: The soundings of the first two seasons. Bibliotheca Mesopotamica 20, 49-56.

WIDELL, M.
2005 Some observations on the administration, agriculture and animal management of Tell Beydar, Ugarit-Forschungen 35, 717–733.

WILKINSON, T.J.
1990 Soil development and early land use in the Jazira region, Upper Mesopotamia, World Archaeology 22.1, 87-103.

WILKINSON, T.J. – TUCKER, D.J.
1995 Settlement development in the north Jazira, Iraq; A study of the archaeological landscape.

5. Micromorphological analyses of selected samples from the depression around Tell Mozan

Paul Goldberg

1. Introduction

The deposits studied here come from a profile that was observed and described in 1999 in a 3.5 m deep pit in the central part of the ditch; it represented a relatively wide (over 100 m), roughly ring-shaped depression around Tell Mozan[1]. The ditch filling is ca. 2 m thick and displays signs of two or three phases of sediment accumulation. The sequence is crowned by a soil (possibly a Chromic Luvisol). A characteristic feature of the pedosediment is represented by several forms of secondary carbonates, which formed obviously *in situ* after sedimentation. The lower part of the profile is represented by an artefact-free loam that is overlain by an 18 cm thick, slightly darker layer that contains dark-colored aggregates (whose origin is problematic: there are fractions lighter than water and heavier) and burned ceramic shards. This layer is unusually uniform in thickness, even in places where it is not strongly horizontal. In the field, it was thought that this may represent a fossil plough horizon; it was thought that it would be exciting to try to test this hypothesis, both in terms of theory and laboratory data.

Samples for micromorphological analyses were collected by K. Pustovoytov in order to:

1. Characterise the pedosediment - its morphology, properties and the chronological framework of its formation; this study would be among the first descriptions of the content of the city ditches of Mesopotamian tells;
2. Interpret these characteristics in terms of the processes of sedimentation and soil formation; and
3. Deduce the significance of the archaeological remains and features, such as the function of the ditch.

2. Methods

Thin sections were collected as undisturbed blocks in the field. They were sent to Thomas Beckman and impregnated and processed into covered thin sections, 60 × 48 mm. The thin sections were examined with binocular and petrographic microscopes in plane-polarised (PPL) and cross-polarised (XPL) light, at magnifications ranging from 20x to 200x.

The descriptive nomenclature used herein follows that of Courty *et al.*[2] and Bullock *et al.*,[3] with the modifications of Stoops.[4]

3. Results

Sample 1 (depth between 17-23 cm)

This sample consists of gravelly and sandy rounded calcareous clasts in a calcareous silty clay matrix (fig. 1). Specifically, the coarse fraction is predominantly comprised of well-rounded sand, granule, and gravel-sized grains of limestone and calcareous elements (i.e. fragments of calcrete or stone pendants). Occurring in very trace amounts are angular grains of quartz, and dark brown clasts of silty clay that appear to represent fragments of either pottery or daub. Some remains of modern roots appear, with some having been eaten by soil microfauna that left behind fine rounded, silt-sized excrements.

---

1 See Deckers and Pustovoytov, this volume chapter 4.

2 Courty *et al.* 1989.

3 Bullock *et al.* 1985.

4 Stoops 2003.

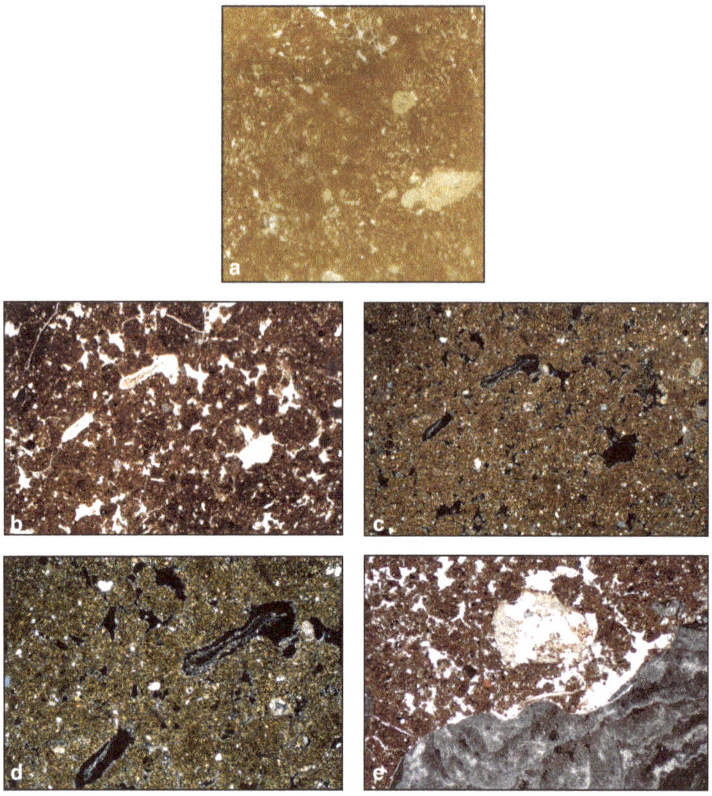

Fig. 1a: Macroscan of sample 1 showing the aggregated nature of the material and its vughy nature. The large grain in the centre right is in fact a rounded clast of soil carbonate (either a piece of calcrete or a pendant) that has been reworked from somewhere possibly upslope. PPL; width of view is 48 mm. 1b. The aggregates shown here result from marked bioturbation, possibly by earthworms or other soil fauna, as well as by roots; the latter is shown by the presence of two elongated voids (channels) in the centre-left part of the photograph. PPL; width of view is 4.7 mm. 1c. Same as in fig. 1b but in XPL. Note the finer, silt-sized quartz within the aggregates. 1d. Detail of root channels shown in fig. 1c. XPL; width of view is 2.35 mm. 1e. Aggregated fine material in the upper part of the photograph with a portion of a carbonate pendant or calcrete fragment in the lower right-hand corner. PPL; width of view is 4.7 mm.

The fine fraction is composed of calcareous clay with inclusions of coarse silt-sized (~50 μm) quartz, muscovite, as well as slightly coarser (vf sand/coarse silt) angular pieces and fragments of charcoal; both of these components are well integrated into the matrix, although the charcoal is slightly coarser and concentrated at the very bottom of the slide. Some phytoliths could be observed, but only on close inspection at relatively high magnifications of 200x. Locally, as can be seen in the macroscan of the thin section, are millimetre-sized domains or possibly former aggregates that are slightly richer in clay, with a slight reduction in quartz silt.

The microstructure varies from moderately separated and weakly separated granular to vughy microstructure; some fissures can be seen in the lower part of the sample. Granules of the fine fraction are effectively soil faunal excrements, and no doubt the biological activity is responsible for the finely comminuted nature of the charcoal and its incorporation into the overall mineral fine fraction.

Comments: Overall this is calcareous silt – likely derived, ultimately, from fluvial materials - which appears to have been subjected to intense biological reworking by soil microfauna. This is shown by the porosity and microstructure, as well as the excremental fabrics.

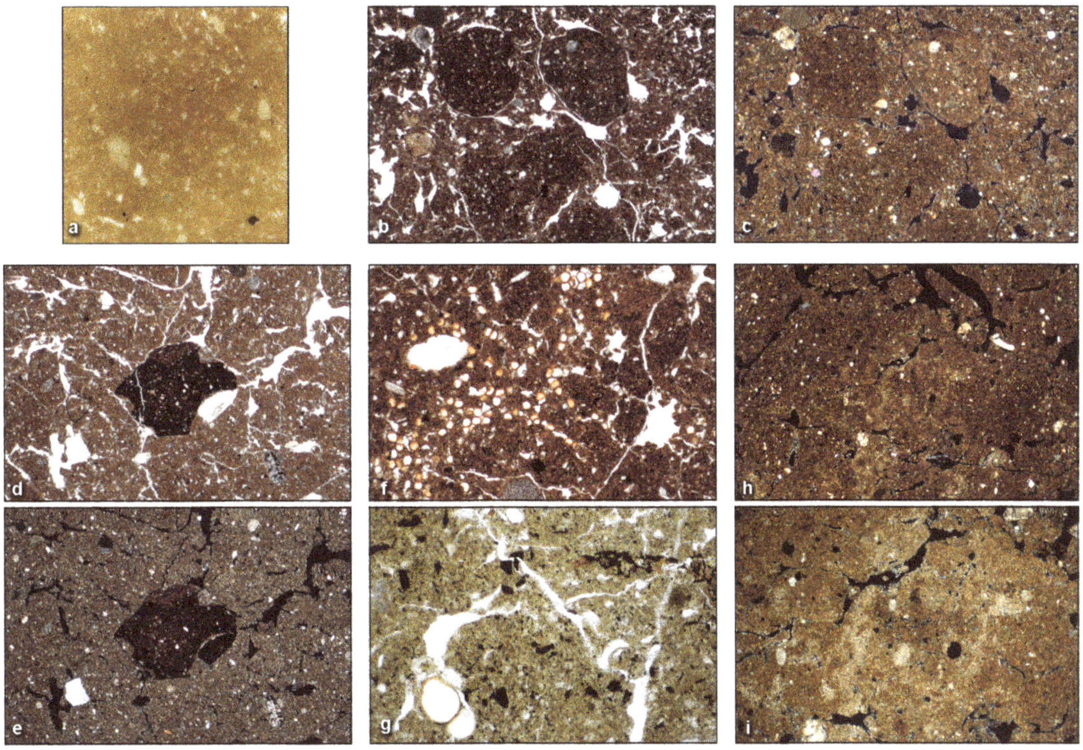

Fig. 2a: Macroscan of sample 2 showing an angular fragment of dark brown daub/brick/pottery in the central-right of the photograph. The open microstructure is generally the same as that of sample 1. PPL; width of view is 48 mm. 2b. Rounded soil aggregates that have been transported or partly affected by some churning or rolling movement. PPL; width of view is 4.7 mm. 2c. Same as in fig. 2b but in XPL. The overall similarity of the matrix material both within the aggregates and in the less differentiated areas is more apparent in this view. Note that the rounded aggregate in the upper left is somewhat clayier than the others, but overall is not significantly different. 2d. Angular clast of dark brown silty clay (possibly representing burned daub, mud brick, or pottery) within a biologically reworked calcareous clay matrix. PPL; width of view is 4.7 mm. 2e. Same as in fig. 2d but in XPL. Note the general similarity of this darker inclusion with that of the surrounding matrix, suggesting that past occupants used the same material for making their adobe, etc. 2f. A concentration of silt-sized spores within the matrix. These are found only within a localised area of the slide. PPL; width of view is 2.35 mm. 2g. Higher magnified view of a single spore in lower left, along with fine fragments of charcoal or organic matter that have been incorporated in to the matrix by bioturbation. PPL; width of view is 960 µm. 2h. Matrix with diffuse carbonate precipitation in what were once possible vughs or fissures. XPL; width of view is 4.7 mm. 2i. Detail of photograph at left with increased illumination from substage condenser, which displays the secondary carbonate more clearly. XPL; width of view is 2.35 mm.

Sample 2 (depth between 24-40 cm)

Overall, this has the same basic components as the first sample, but it differs somewhat in its proportions, structure, and organisation. Firstly, in terms of content, it contains a slight increase in size of the anthropogenic material, namely clasts of dark brown silty clay (representing adobe, daub, or pottery); an isolated, rounded sand-sized bone fragment was also observed (fig. 2). Although the microstructure is not radically different, we begin to see a greater development of planes and fissures here, with a slight suggestion of angular peds. Furthermore, we can see the presence of well-rounded grains of fine material here that attain 1-2 mm in diameter. Although they have essentially the same composition as the rest of the sample, they are easily differentiated from the bulk of the matrix. Fine charcoal and phytoliths are equally mixed into the matrix here as above. Concentrations of silt-sized spores were observed in localised areas of the slide. Finally, we note the appearance of weakly developed impregnations of calcium carbonate in what appear to have been former fissures; incipient carbonate hypocoatings can also be observed.

Comments: Although of similar material, this sample shows a greater degree of pedogenetic development in the form of fissured microstructure and incipient carbonate precipitation. Also noteworthy is the

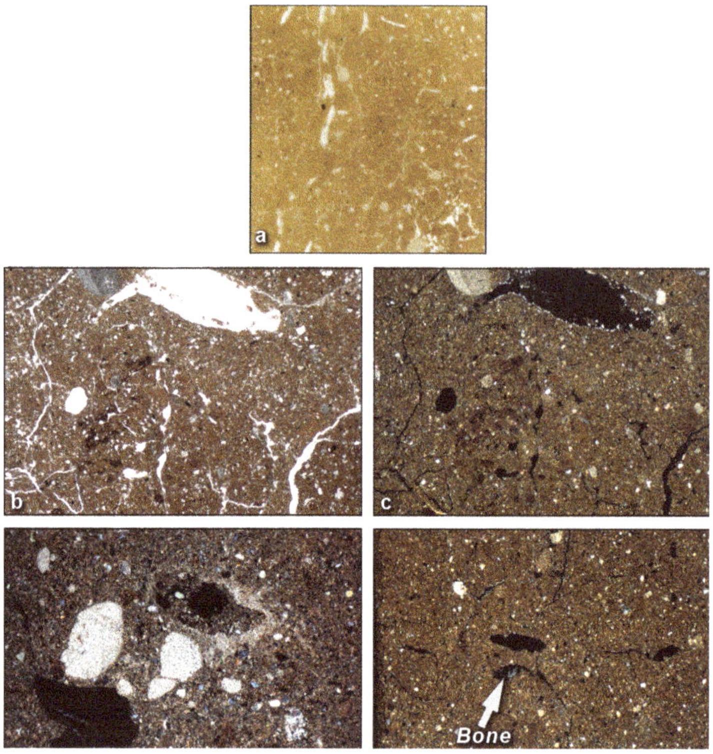

Fig. 3a: Macroscan of sample 3. Note the presence, as elsewhere, of rounded granules of limestone scattered throughout the slide. The darker grains are well-rounded silty clay aggregates of soil, ceramic, adobe, or daub. They are generally quite well-rounded, suggesting that they were transported over a relatively long distance and do not represent colluvial reworking of these materials from an adjacent source. The vertical channels in the centre of the slide are formed from a modern root, traces of which are still visible. PPL; width of view is 48 mm. 3b. The microstructure in this view is somewhat more sub-angular blocky, although biological activity is quite marked, as shown by the concentrically-shaped passage feature in the left third of the photograph. PPL; width of view is 4.7 mm. 3c. Same as in fig. 3b but in XPL. Both the passage feature and the sub-angular blocky structure are somewhat more distinct in this view. 3d. Incipient hypocoating in the void at the upper part of the photograph. XPL; width of view is 2.35 mm. 3e. Bone fragment (indicated) can be seen in the middle part of the photograph. This is one of the only bone fragments in the slide; overall bone fragments are rare in all the thin sections. XPL; width of view is 4.7 mm.

appearance of bone in this sample and a larger piece of ceramic or adobe-like material, which may signal an increased input of anthropogenic components at this time.

Sample 3 (depth between 40-70 cm)

This is generally the same material as observed in the sample above it but with a few notable changes. We note firstly, an increase in rounded sand-sized grains of reddish brown and dark brown silty clay, which is probably adobe, daub or ceramic; the small size of these grains precludes definitive assessment (fig. 3). Biological activity is still quite prominent and numerous passage features could be observed. Such activity is also likely responsible for localised domains (~1 mm in diameter) of silt concentrations where some of the fine silt and clay are not present. The microstructure here also is more subangular blocky, although many intrapedal vughs, vesicles, and channels are present; the channels, in particular, appear to be the result of modern roots. Some rounded, slightly clayier matrix aggregates are present, as in sample 2, but these are less pronounced and smaller than in the previous sample. Finally, carbonate precipitation is about the same as seen in the preceding sample, but some void hypocoatings begin to appear.

Comments: In sum, this is similar to the overlying materials, but the more pronounced subangular blocky structure suggests a more mature soil-like material. As in the overlying sample, the number of rounded grains of adobe, and other such material increases, possibly representing increased dumping activity or transport by the ditch. As seen previously, their well-rounded nature calls for relatively longer transport distances.

Sample 4 (depth between 120-145)

The dark, rounded, silty clay clasts are noticeably more abundant in this sample as can readily be seen in the macroscan that follows (fig. 4). These grains are not only more abundant in this sample but also vary in color from bright yellowish brown (resembling soil material) to darker brown. Some contain phytoliths as well as elongated voids that appear to be pseudomorphs of vegetation, similar to those found in mud bricks in the Middle East; phytoliths also occur scattered throughout the matrix. Some of these grains have isotropic domains, which are reminiscent of heated mud brick material or pottery.[5] Furthermore, in the bottom quarter of the slide, pieces and flecks of fine charcoal could be seen to be well-worked into the matrix, generally by bioturbation. The porosity pattern here is much more open, with numerous vughs, channels, vesicles, and some fissures. The latter are less prominent than in overlying samples and the tendency here toward a subangular blocky microstructure – evident in the overlying sample 3 – is scarcely visible.

Comments: This sample marks a notable increase in the abundance of anthropogenic components, particularly silty clay aggregates, some of which appear to be pieces of mud brick containing phytoliths; the latter are overall more abundant here than in the overlying samples. In addition, some bone fragments were also observed. The sample would seem to indicate either a greater increase in human activity, or at least the activities were closer to where the sample was taken. Finally, the change in microstructure here to a more crumbly nature suggests an increase in biological activity and disturbance. It may also be associated with increased human activity, as seen here by the rise in anthropogenic components, or then perhaps attributed to a greater time of exposure at the surface, which allowed for a greater degree of bioturbation.

Sample 5 (depth between 155-180 cm)

This sample is more compact than the one above it, and the subangular blocky microstructure reflects this appearance (fig. 5). Nevertheless, excremental features – with localised granular aggregates of matrix – can also be observed. The matrix is somewhat finer and clayier than many samples and exhibits localised domains that are more orange and clayey than the surrounding somewhat siltier ones. In this sense, the situation is similar to that of sample 2, but here there are fewer rounded aggregates than in the sample immediately above. Finely comminuted charcoal is present and finely mixed into the groundmass by bioturbation. In addition, small (~100 µm in diameter) areas are stained with what appears to be Fe/Mn. Interestingly, these localised 'spots' appear to pre-date the formation of secondary carbonate precipitation, which here is better developed than in any of the overlying samples; it is expressed as void hypocoatings and incipient nodules. Finally, we note the occurrence of a millimetre-sized grain comprised of calcitic spherulites mixed with quartz silt and some clay; the spherulites, noted by their pseudo uniaxial figures, are commonly interpreted to be the remains of dung. Interestingly, an elongated phytolith (grass?) was found within this sand-sized grain.

Comments: This sample differs from the one above it in a few ways. To start with, it would appear to contain some cultural material in the form of charcoal (albeit biologically reworked) and dung. Next, it demonstrates the occurrence of some Fe/Mn staining, whose significance is unclear but could be related to the possible effects of organic matter or wetter soil conditions. Also, the developmental degree of secondarily precipitated carbonate is greater here, with incipient nodule formation as well as carbonate hypocoatings. As a final point, we observe a return to a blockier microstructure in comparison to the open, vughy one of the immediately overlying sample (see fig. 9); this may be a possible reflection of burial depth.

---

[5] see Courty *et al.* 1989.

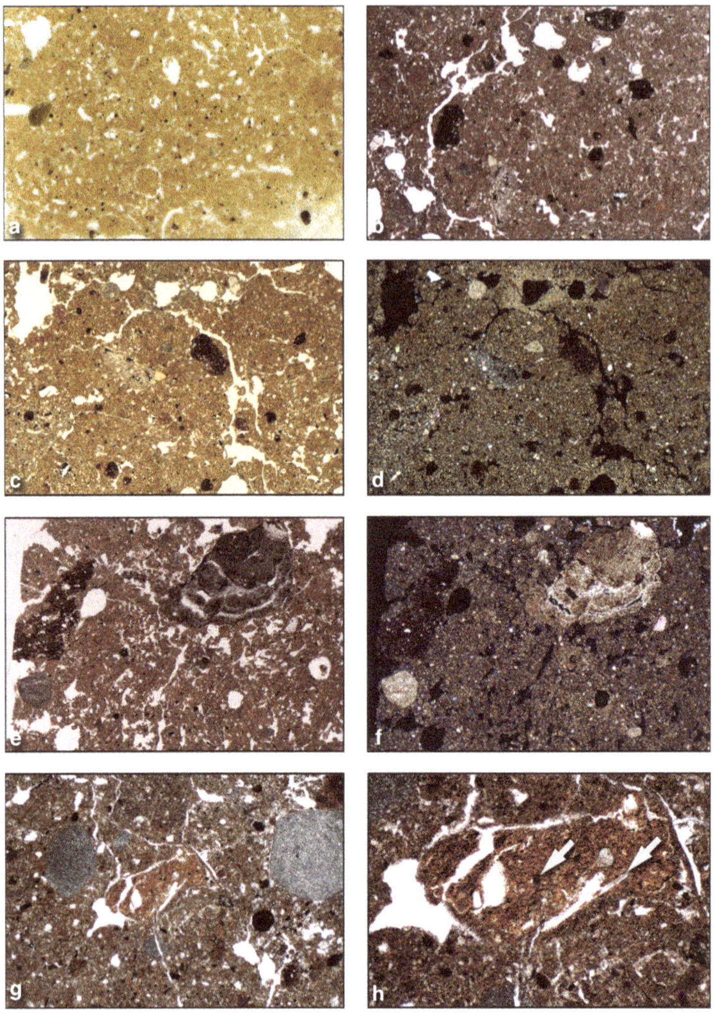

Fig. 4a: Macroscan of sample 4, showing porosity and increase in darker brown 'flecks' as compared with overlying samples. These are rounded clasts of silty clayey material that represent adobe, daub or pottery. PPL; width of view is 48 mm. 4b. Silty clayey grains within a porous, bioturbated matrix. PPL; width of view is 4.7 mm. 4c. Clayey clasts mixed into a porous, bioturbated, calcareous clayey fine fraction. The voids in the upper central part of the photograph exhibit carbonate hypocoatings. PPL; width of view is 4.7 mm. 4d. Same as in fig. 4c but in XPL. The carbonate hypocoatings at the top are more evident in this view. 4e. A large, angular adobe or daub-like fragment can be seen at the top; in the lower right is an angular clast of bedded calcium carbonate similar to a carbonate pendant that forms on the under sides of rocks in this part of the world. The porous, bioturbated nature of the deposit is clearly visible here and differs slightly from the more fissured overlying sample 3. PPL; length of view is 4.7 mm. 4f. Same as in fig. 4e but in XPL. Note the silt fraction within the clay and overall moderate to poor sorting. 4g. Reddish brown clayey aggregate (adobe?) in centre containing phytoliths. PPL; width of view is 2.35 mm. 4h. Detail of fig. 4g with two phytoliths. Their occurrence along with the void structure is reminiscent of adobe bricks tempered with straw, which are widespread over this geographic region. PPL; width of view is 960 μm.

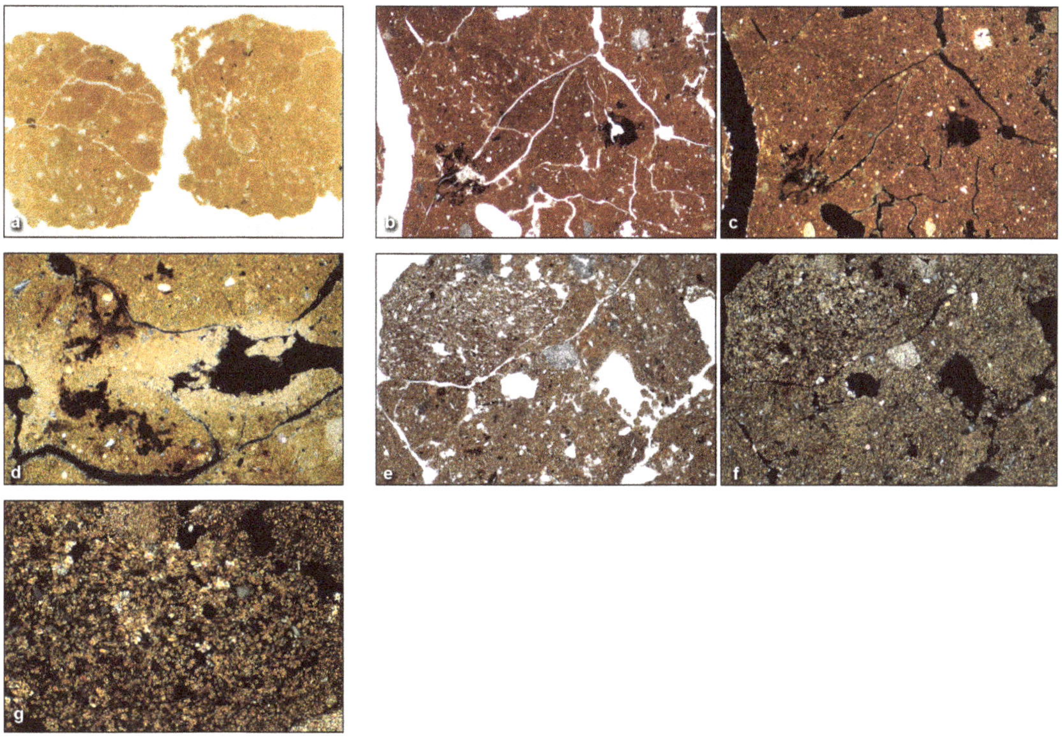

Fig. 5a: Macroscan of two peds from sample 5. Note the fissured microstructure with subangular blocky peds, as well as some intrapedal vughs. PPL; width of view is 48 mm. 5b. Photomicrograph showing overall uniform, compact, calcareous clayey matrix with many fissures locally producing a sub-angular blocky microstructure. Note also the subtle changes in colour, with the more orange areas being somewhat richer in clay but largely lacking in quartz silt; the splotchy dark brown/black spots appear to be Fe/Mn impregnations. PPL; width of view is 4.7 mm. 5c. Same as in fig. 5b but in XPL. 5d. Light-coloured carbonate hypocoating around void in the centre, which at left appears to displace and therefore post-date the dark irregular Fe/Mn impregnations. XPL; width of view is 2.35 mm. 5e. Detail of topmost part of the thin section, illustrating in the upper left hand corner an aggregate of what appears to be dung as shown by the presence of spherulites (fig. 5f). PPL; width of view is 2.35 mm. 5f. Same as in fig. 5e but in XPL. Note the slightly coarser and brighter appearance of the dung fragment in the upper left. 5g. Detail of dung fragment showing spherulites with characteristic pseudo uniaxial shape in XPL; width of view is 960 μm.

Sample 6 (between 180-197 cm)

This is a relatively poorly sorted deposit in which we see a return to numerous anthropogenic inclusions in the form of charred vegetal remains (some finely comminuted), sand, and granule-sized pieces of brown silty clay (fig. 6). Some of these pieces exhibit channel voids which are pseudomorphic after plant remains and suggestive of adobe or daub.[6] Although much of the mass of material in the slide is loose and not compacted (especially with respect to the overlying sample), some fissures and more compacted domains are visible, particularly in the lower part of the slide. These grains vary in color from lighter to darker brown, and some contain phytoliths and charred material, and others do not. Carbonate hypocoatings are thicker and larger in diameter than in overlying samples.

Comments: This sample shows marked anthropogenic inputs of likely adobe, as well as charred remains that were subjected to modest biological reworking and secondary cementation by carbonates. The increased degree of development of carbonate hypocoatings than in the overlying sample could indicate a greater depth of burial, age, or moisture conditions. It seems likely that the differences in color of these aggregates, from lighter to darker brown, is a partial function of their organic content and their possible degree of heating.

[6] Courty *et al.* 1989.

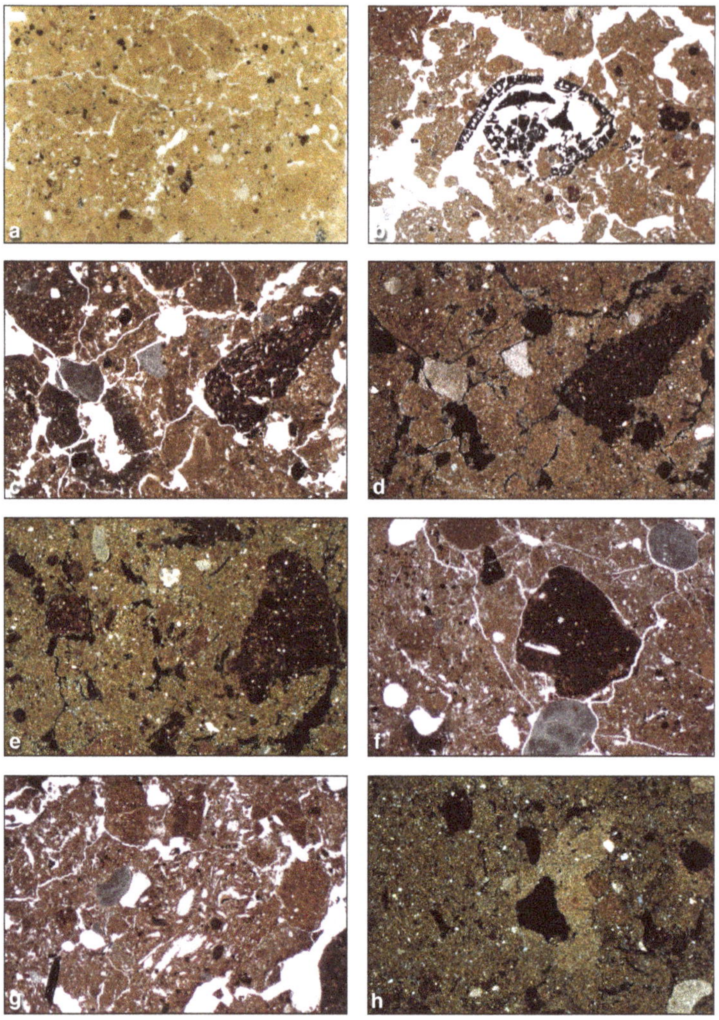

Fig. 6a: Macroscan of sample 6 showing numerous darker brown grains of silty clay (soil aggregates, adobe and/or daub, of pottery). PPL; width of view is 48 mm. 6b. Grain of charcoal within a porous, biologically reworked, calcareous silty clay matrix. PPL; width of view is 4.7 mm. 6c. Piece of dark brown silty clay at the right in a crumbly calcareous clay. Note the different types of voids, consisting of fissures, vesicles (rounded voids) and chambers. The large sub-rectangular void in the lower left displays a carbonate hypocoating as shown by its somewhat greyer color. PPL; width of view is 4.7 mm. 6d. Same as in fig. 6c but in XPL. 6e. Several sizes of angular, dark brown silty clay aggregates. XPL; width of view is 4.7 mm. 6f. Dark brown silty clay granule-sized grain, showing elongated void with scant remains of vegetation; this type of void is typical of adobe or daub. PPL; width of view is 4.7 mm. 6g. Mixture of silty clayey aggregates. Note the large rounded grain in the lower half of the photograph showing numerous channels representing plant pseudomorphs as typically found in adobe or daub. PPL; width of view is 4.7 mm. 6h. Carbonate hypocoating is shown around the void in the centre. These types of features are thicker and better developed here than in the overlying deposits. XPL; width of view is 4.7 mm.

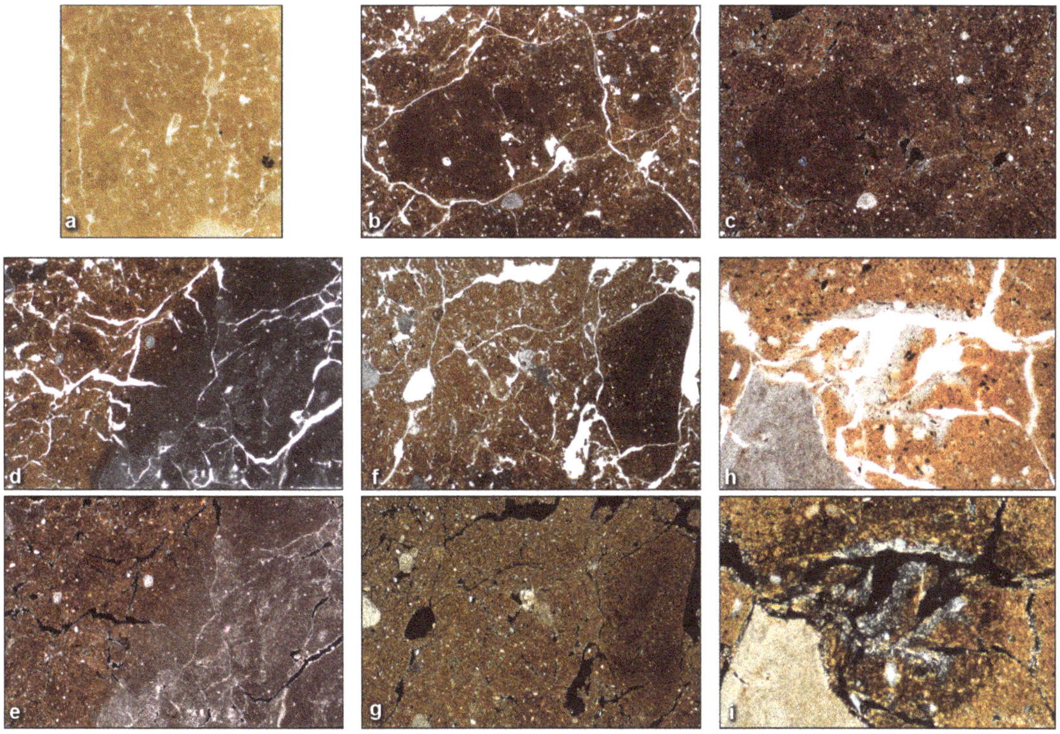

Fig. 7a: Macroscan of sample 7. The fissures and blockier microstructure are more prominent here and contrast with the more granular porosity of the overlying sample (see also fig. 9). Note the cm-sized rounded carbonate nodule at the very bottom of the section. PPL; width of view is 48 mm. 7b. Chunky fragments of slaking crust are visible in the centre of the photograph within an angular blocky ped. PPL; width of view is 4.7 mm. 7c. Same as in fig. 7b but in XPL. The bedding within the clast in the centre is more evident in this view. 7d. Carbonate nodule at right grades into calcareous silty clayey matrix. PPL; width of view is 4.7 mm. 7e. Same as in fig. 7d but in XPL. 7f. A reworked clast of a slaking crust appears at the right. PPL; width of view is 4.7 mm. 7g. Same as in fig. 7f but in XPL. 7h. An irregular void showing slight 'bleaching' by removal of iron. This is not a textural pedofeature as the fine fraction is the same across the color boundary. PPL; width of view is 2.35 mm. 7i. Same as in fig. 7h but in XPL. The similarity in texture of the gray and brownish areas shows that this is a chemical difference and not a difference in the make-up of the fine fraction.

Sample 7 (depth between 197-210 cm)

This sample shows a marked decrease in anthropogenic components in comparison to the overlying sample and, for example, is similar to deposits found in sample 2. The composition is calcareous silty clay but with grains and domains of clayier and more orange clay; rounded carbonate grains also occur in similar proportions as in other samples (fig. 7). Some of these clayey domains show traces of bedding and would appear to be surface slaking crusts (or fine alluvium) that have been incorporated into the deposit. Dark brown silty clay grains are sparsely scattered throughout the matrix and are relatively rare; some contain elongated voids and thus suggest that they are adobe and/or daub. Pieces of organic matter and/or charcoal are less than in the sample immediately above, and are very finely divided and dispersed into the groundmass. Additionally, Fe/Mn staining is represented by small (~ 250 µm) specks. Secondary carbonates are more pervasive here and occur roughly as centimetre-sized nodules with only some void hypocoatings. Finally, we note the presence of traces of iron depletion as can be seen in one of the voids. This has not been observed elsewhere.

Comments: This sample is much less anthropogenic but more pedogenic in character, as shown by the larger and better developed carbonate nodules and widespread impregnation of carbonate. In addition, the iron depletion feature, while uncommon, suggests some stagnation of water on the surface. It may also be related to the greater amount of clay and slaked clay clasts in this sample, which in turn would point to some surface crusting associated with water.

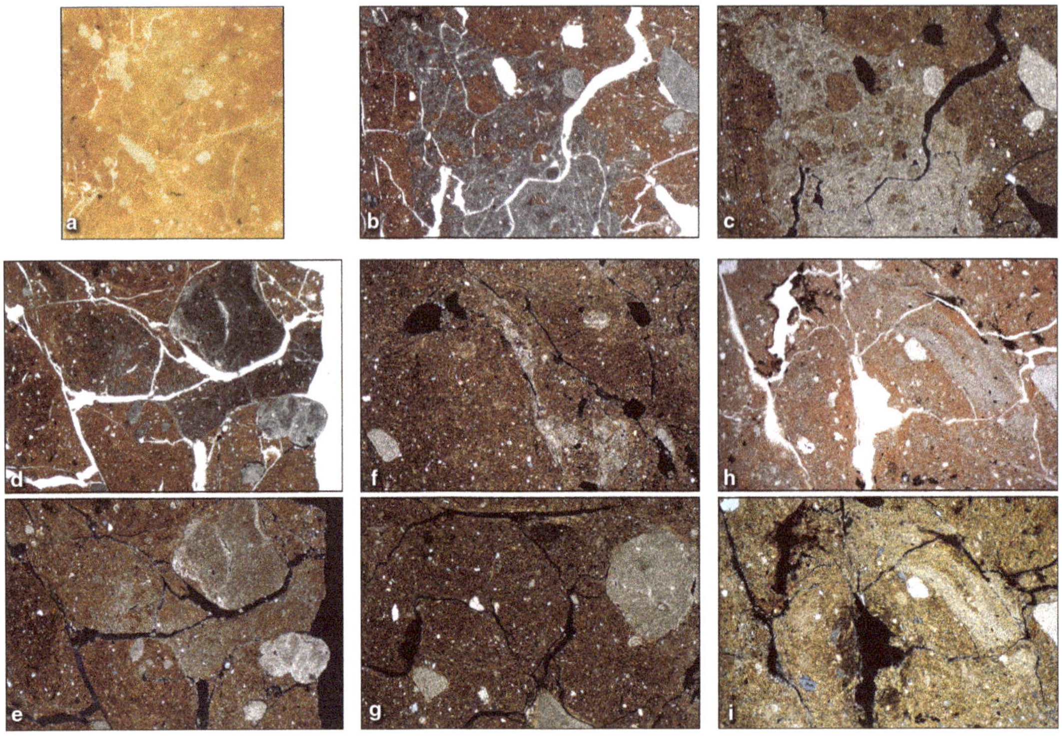

Fig. 8a: Macroscan of sample 8. The lighter domains of carbonate impregnation are clearly evident in this scan. PPL; width of view is 48 mm. 8b. Complex carbonate nodule showing carbonate engulfing the silty clayey matrix. PPL; width of view is 4.7 mm. 8c. Same as in fig. 8b but in XPL. Note the overall compacted and slightly clayier nature of the material as compared to the overlying deposits; the deposit is also better sorted as a whole. 8d. Better defined, rounded carbonate nodule than present in fig. 8b and 8c. Note the many fissures and cracks. PPL; width of view is 4.7 mm. 8e. Same as in fig. 8d but in XPL. 8f. Carbonate precipitated as ill-defined nodule at base, and along sub-vertical fissure in the centre of the photograph. A carbonate hypocoating is also visible around the circular void in the upper right-hand part of the photograph. PPL; width of view is 4.7 mm. 8g. Partly vertic nature of this part of the deposit is shown by the notable orientation of clay, including granostriated b-fabric around the carbonate clasts and porostriated b-fabric along the horizontal plane in the upper part of the photograph. XPL; width of view is 4.7 mm. 8h. Detail of sample showing Fe/Mn staining along voids. In the upper right-hand part the Fe/Mn appears along a fissure and post-dates the phase of carbonate precipitation. PPL; width of view is 2.35 mm. 8i. Same as in fig. 8h but in XPL. The secondary carbonate in the right-hand part of the photograph is more visible in this view. Also note the non-calcareous nature of the dusty reddish clay around the void in the centre.

Sample 8 (depth between 210-230 cm)

This lowermost sample is strongly impregnated with carbonate, as can be clearly seen in the field and in the macroscan (fig. 8). The carbonate occurs as nodules but, more typically, as impregnated zones along fissures and cracks, in which the carbonate engulfs the silty clay matrix. The fine fraction in this sample appears somewhat clayier than in the overlying deposits and overall, the material tends to be compact with relatively few biologically produced voids or aggregates; the microstructure is subangular blocky with many fissures and cracks, along with some intrapedal voids.

Fe/Mn staining is visible and its precipitation is largely penecontemporaneous with the carbonate, as it can be seen to line fissures developed in carbonate domains but is also slightly incorporated into some of the carbonate-rich areas. On the whole, however, Fe/Mn precipitation appears to be a late stage effect and the majority of it occurs along the most recent set of fissures and cracks. Finally, we note the presence of some vertic aspects of this layer as shown by granostriated and porostriated birefringence fabrics (b-fabrics), representing preferred orientation of the clayey matrix. The mild shrink-swell results in formation of the fissures, which were favorable venues for carbonate precipitation.

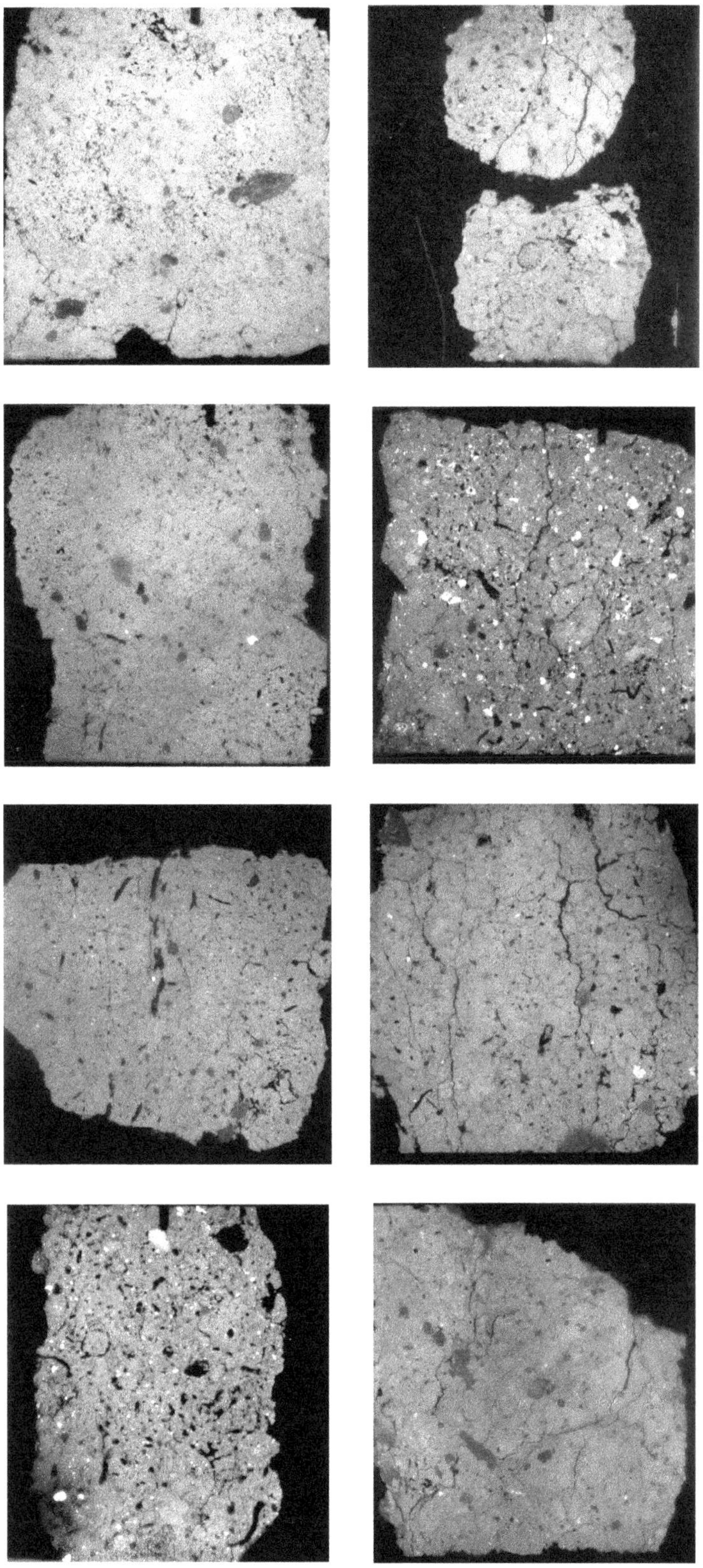

Fig. 9: Contact sheet of black and white thin section macroscans arranged from top (upper left) to bottom (lower right). The original color scans (see preceding samples) were manipulated with Adobe® Photoshop® by converting them to greyscale images and then inverting them in order to bring out the porosity patterns in black.

Comments: This sample displays the most well-developed soil formation features than in any of the samples, including more massive and larger carbonate precipitation, Fe/Mn precipitation along voids, and the occurrence of b-fabrics. This sample is clayier than others and shows the best sorting, with relatively little coarse silt. Its sterile nature, as seen in the field, is confirmed in thin section; it would also seem that this is a fossil soil.

4. General Comments

In looking at the thin sections from the profile en masse, the following comments can be made:

1. The sediments are composed of silty clay and clay with rounded granule-sized grains of limestone and reworked carbonate pendants and calcrete fragments. The textures are variable from top to bottom, but nevertheless, the lowermost deposits are better sorted and overall finer grained. These deposits appear ultimately to be alluvium mixed with some colluvial material, possibly derived from the tell.

2. Certain samples (e.g. 180-197) are relatively rich in human-derived materials. These take the form of yellow brown or dark brown silty clay, that represents adobe (as shown by elongated voids after plant fragments), daub, and/or possibly ceramic. Bone fragments are generally quite rare, as is charcoal. The latter, however, occurs as very fine, silt-sized fragments that have been comminuted and reduced in size by biological burrowing activity. The origin of these anthropogenic components is not clear, but they could represent either short-lived occupations, dumped material, or objects that have been washed down from the tell.

3. Soil formation processes are visible throughout. These can be seen in the biological activity and extensive burrowing, and also in the precipitation of carbonate. The latter occurs in most sections but, as in the field, becomes better expressed and more prominent with depth; with the deepest sample likely being a buried soil without cultural inputs. In addition, the lowermost slide shows some features seen in vertisols (e.g. granostriated and porostriated b-fabrics).

4. Additionally, changes in soil formation processes are visible in the porosity patterns, which appear to point to former surface horizons. Fig. 9 shows black and white images of scans of entire thin sections. Although fissured microstructures are present in many of the samples, they are particularly well-developed in samples from 155-180, and the bottom two samples, 197-210 and 210-230. On the other hand, granular fabrics resulting from bioturbation are visible in most samples (except for the bottommost), but are particularly prominent in samples 120-145 and 180-197. These latter two depths would appear to point to still-stands of sedimentation, allowing for extensive reworking by soil fauna.

Bibliography

BULLOCK, P. – FEDOROFF, N. – JONGERIUS, A., – STOOPS, G., – TURSINA, T.
1985 Handbook for Soil Thin Section Description.

COURTY, M.-A. – GOLDBERG, P. – MACPHAIL, R.I.
1989 Soils and Micromorphology in Archaeology.

STOOPS, G.
2003 Guidelines for Analysis and Description of Soil and Regolith Thin Sections.

6. Riverine development through time in the Tell Leilan surroundings

Katleen Deckers

1. Introduction

The earliest evidence for occupation at Leilan dates back to approximately 5000 BC and is mainly concentrated on the "citadel", i.e. the main mound (measuring fifteen hectares). By about 2600 BC, it grew into a large city of roughly ninety hectares. At this time a lower town, with an associated city wall and ditch, was established.[1] Later, at around 2300 BC, the site came under Akkadian control and some reorganisation took place.[2] Shortly after, by about 2200 BC, like many other sites in the area, Leilan was abandoned for a few hundred years.[3] After the conquest of the region by the Assyrian king Shamshi-Adad I (1813–1781 BC), the site of Tell Leilan (renamed Shubat-Enlil) was revived again and it became capital of his northern Mesopotamian kingdom. In the city a royal palace was built and a temple acropolis to which a straight paved street led from the city gate. There also was a planned residential area and renewed construction works took place on the city wall.[4] The city prospered until King Samsu-Iluna of Babylon sacked it in 1726 BC. Shubat-Enlil never regained its power afterwards. It subsequently became a provincial Hurrian town in the kingdom of Mitanni.[5]

Today, Tell Leilan can be found in the Upper Khabur Basin of northeastern Syria, roughly 500 m south of the confluence of the Wadi Jarrah with the Wadi Siblah and less than two kilometres north of the confluence of the Wadi Jarrah with the Wadi Qatrani. Overall, the site sits between the Wadi Jaghjagh and Wadi Qatrani, both of which directly contour the site (fig. 1). Hence, the rivers must have played a major role in the site's location and economy. Therefore, it is important to gain insight into their evolution during the site's occupation. Moreover, gaining an understanding into the development of the Jarrah is also of importance in reconstructing other sites in their environments, since sites within this region are typically set near to rivers.[6] Presently, the Wadi Jarrah and Qatrani run dry at the height of Tell Leilan but this is due to damming in northern Syria, which is clearly visible on satellite images. What is less obvious from satellite images is why the Wadi Siblah no longer contains perennial water. Today, all three wadis have a meandering morphology with fine-grained sediments within their river beds. The present day vegetation is only represented by non-woody plants. However, it is important to note that up until now at Leilan, from the strata dated to 2300-2200 BC, 70% of the charcoal fragments identified are riverine gallery forest taxa, represented by *Fraxinus*, *Populus/Salix*, and *Ulmus*.[7] This fits well with what is known of the parallel-flowing Wadi Jaghjagh. The latter wadi had a riverine gallery forest growing along its course in the mid fourth millennium BC.[8] Besides the riverine vegetation, park-woodland species are present as well within the Tell Leilan samples. In the Early Bronze Age these samples are mainly represented by oak park-woodland.

Besonen and Cremaschi[9] undertook fluvial geomorphological research in collaboration with the Tell Leilan project. They investigated a 3.4 m deep bulldozer cut in the Jarrah channel, about five kilometres north of Tell Leilan, and found indications for changes in river morphology throughout the Holocene, from meandering to braided and again to a meandering system. Nevertheless, the chronology for these changes is, as yet, undefined.

Based on the aforementioned chronological difficulties, new geomorphological research was undertaken in order to gain insight into the chronology of the fluvial history in the Leilan surroundings. By the summer of 2002 a few natural exposures surrounding Leilan were investigated as part of a regional

[1] Risvet 2005.
[2] Weiss *et al.* 1993.
[3] Weiss *et al.* 1993; Staubwasser – Weiss 2006.
[4] Risvet 2005.
[5] Weiss *et al.* 1993.
[6] Deckers – Riehl 2008.
[7] Deckers – Pessin, this volume chapter 3.
[8] Deckers – Riehl 2007.
[9] 2002.

geoarchaeological survey in connection with the Tell Mozan excavations. Additionally, in June 2006, a more detailed field study was undertaken as part of the Tell Leilan excavations.

2. Methods

In order to gain a preliminary understanding of the landscape changes, CORONA satellite images by the mission D025-012 1102-1 AFT 11 Dec were investigated and overlain by exaggerated 3D-SRTM data. Furthermore, selected sequences were studied in the field, and sediments and soils were described following standard methodology. The chronology of the sequences is based on radiocarbon dating (tab. 1) of shells and OSL dating. OSL samples were retrieved at night under total dark conditions on sections that were freshly cleaned under a black plastic bag and investigated by Dr. Dimitri Vandenberghe[10] and Dr. Alexandra Hilgers.[11] Radiocarbon dates were performed by the Kiel laboratories under the direction of Prof. Piet Grootes.

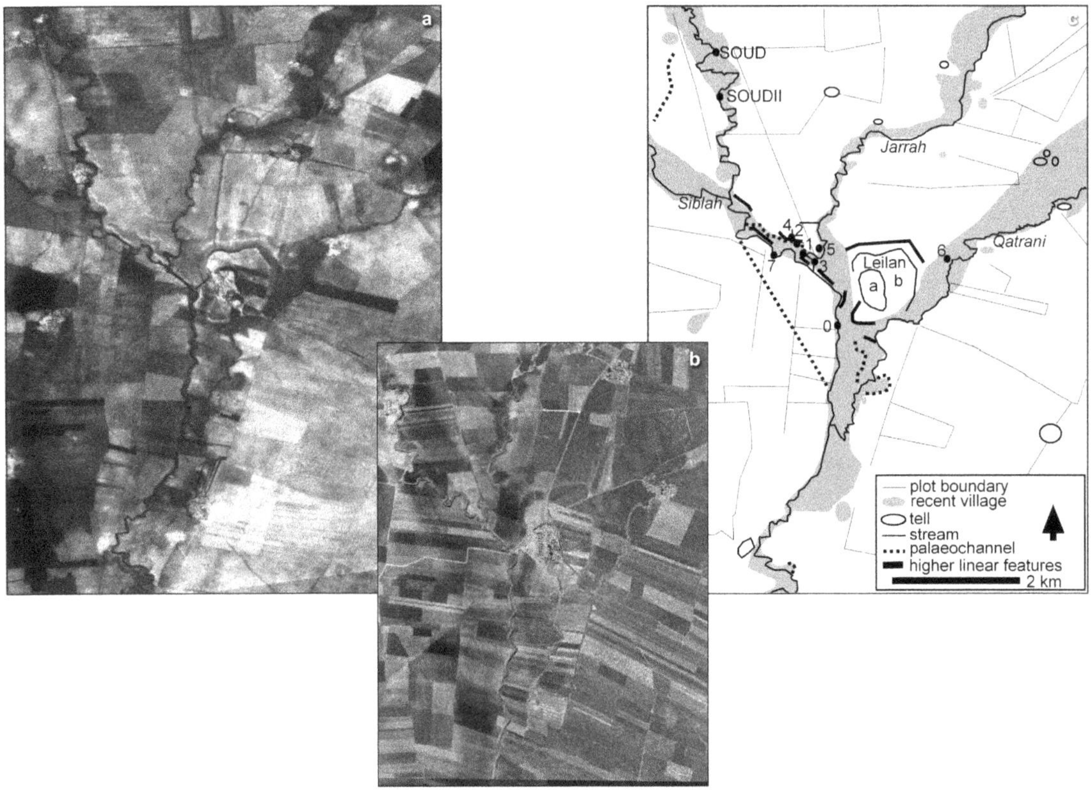

Fig. 1: a. Corona satellite image of the Tell Leilan surroundings (mission D025-012 1102-1 AFT 11 Dec); b. Google Earth image of the same area (shading anomalies due to multiple images being stitched together); c. map indicating the examined profiles, based on information from Google Earth and Corona satellite images. LEI profiles are indicated with Arabic instead of Roman numerals.

Table 1: Radiocarbon dates from the profiles studied in the Tell Leilan region.

Sample number	Material	Radiocarbon age
KIA 30900	bivalve	2030 ± 30 B.P.
KIA 30899	*Unio* shell	5100 ± 30 B.P.
KIA 30898	*Corbicula fluminalis* shell	805 ± 20 B.P.
KIA 30897	*Corbicula fluminalis* shell	> 1954 AD

[10] Vandenberghe, this volume chapter 10.

[11] Hilgers, this volume chapter 11.

3. Results

3.1 Wadi Siblah

Satellite image analysis for the Wadi Siblah showed some interesting features about the riverine history in the neighbourhood of Leilan. One important element is the presence of linear levees, which are indicative of canalisation (fig. 1). They are not connected with the present day course, as they were breached at several locations by recent meanders and also were not on both sides of the current river. The levee knicks ninety degrees, to course parallel and directly adjacent to the town wall. There is no natural obstruction that could have caused this knick. Hence, all evidence seems to indicate canalisation of the course at the time of construction or use of the lower town wall. Lastly, the feature cannot be traced back over a larger distance.

In between these levees a meandering palaeochannel is visible on the satellite images. It is assumed that this is a later feature than the levees since the linear levees do not indicate a meandering stream. The palaeochannel is marked by darker meandering lines which may be related to a higher moisture content in the sediments. This may be related to the depression that is still visible within the present day landscape.

Two profiles were studied within this palaeochannel: LEI I and LEI II (fig. 2). Both were dug about 3.4 m deep with a bulldozer. They ultimately represent the same sequence, as they are composed of fine-grained material with only few visible stratigraphic changes. The lower twenty centimetres of both sequences consists of sandy material that probably represents palaeochannel deposits. This section is followed by clayey silt/silty clay deposits that contains mudballs (rip-up clasts) and is fining upwards. It contains fragments of *Unio* shells and an undetermined form of land snail. These sediments indicate that this course only contained water intermittently and that the main course may have changed its location. Only in times of flooding were vertical accretion sediments deposited. An OSL date indicates that this channel was abandoned, at the earliest, at about 2.8 ± 0.2 kyr (GLL 060105).[12] A radiocarbon date on an *Unio* shell indicates it was more likely around, or somewhat later than (taking into account the hard water effect), 41 ± 42 calibrated BC (KIA 30900: 2030 ± 30 BP). Soil formation exists within the present day surface, in the form of calcium carbonate accumulations and rubefication. Additionally, carbonate precipitation reaches somewhat deeper in LEI II than LEI I. The difference may be moisture related.

Sequence LEI IV (figs. 1 and 2) is located within a gully that leads to the palaeochannel represented by LEI I and LEI II. This gully is connected with a darker linear feature that could be a hollow way and less likely a palaeochannel. As a whole, the 3.2 m deep sequence consists of clayey silt. Within the lower part, between 2.3 and three metres, mudballs (rip-up clasts) and land snails were present. OSL analysis provided two Pleistocene dates for the lowest metre of sediments within the section (samples C-L2016 and C-L2015),[13] while at roughly two metres from the surface a 2.57 ± 0.29 kyr BP date (C-L2017)[14] exists. However, there is no strong evidence for soil formation or an unconformity between the sediments with Pleistocene and those with Holocene dates. The 2.57 ± 0.29 kyr BP date fits well with the date of the LEI I fill. Moreover, large calcium carbonate flecks at a level of about two metres indicate a considerable phase of soil formation took place since these sediments were deposited.

The present day Wadi Siblah dry riverbed consists of fine-grained material and contains many *Corbicula fluminalis* and other shells that are indicative of perennial water flow from some time ago.[15] Two *Corbicula fluminalis* shells were dated in order to gain some insight into when the river came to occupy its present day course. One mollusc was clearly recent as it contained post-bomb ^{14}C, while the other shell (KIA 30898) was dated to 805 ± 20 uncalibrated BP. Although we must take into account the hard water effect, it is rather unlikely that this is a recent shell. The hard water effect would need to be quite large in order to diminish the effect of the bomb on the date. Hence, the date of the shell suggests that this course or approximately this course – in case the shell was eroded from adjacent, earlier deposited wadi sediments – was also flowing sometime before the early fifties.

Within a present day meander of the Wadi Siblah, profile LEI VII was studied within an eroded concave bank that was more than six metres deep (figs. 1 and 3). Only the lower 2.8 m was naturally exposed and available for study. The bottom levels consist of gleyed clayey sand with many manganese stains, especially in the lower part of the profile. The clayey sand deposits are superimposed by gravel (with clasts less than 1.5 cm) in a clayey sand matrix that is fining upwards to clayey sand. This clayey sand, in turn, is followed again by gravel (containing clasts less than 1.5 cm) in a clayey sand matrix that fines upwards to clayey silt. Again, gravel was deposited on top of this and followed by silty clay. Above this section, a twenty

[12] Vandenberghe, this volume chapter 10.

[13] Hilgers, this volume chapter 11.

[14] Hilgers, this volume chapter 11.

[15] Meijer – Preece 2000, 243.

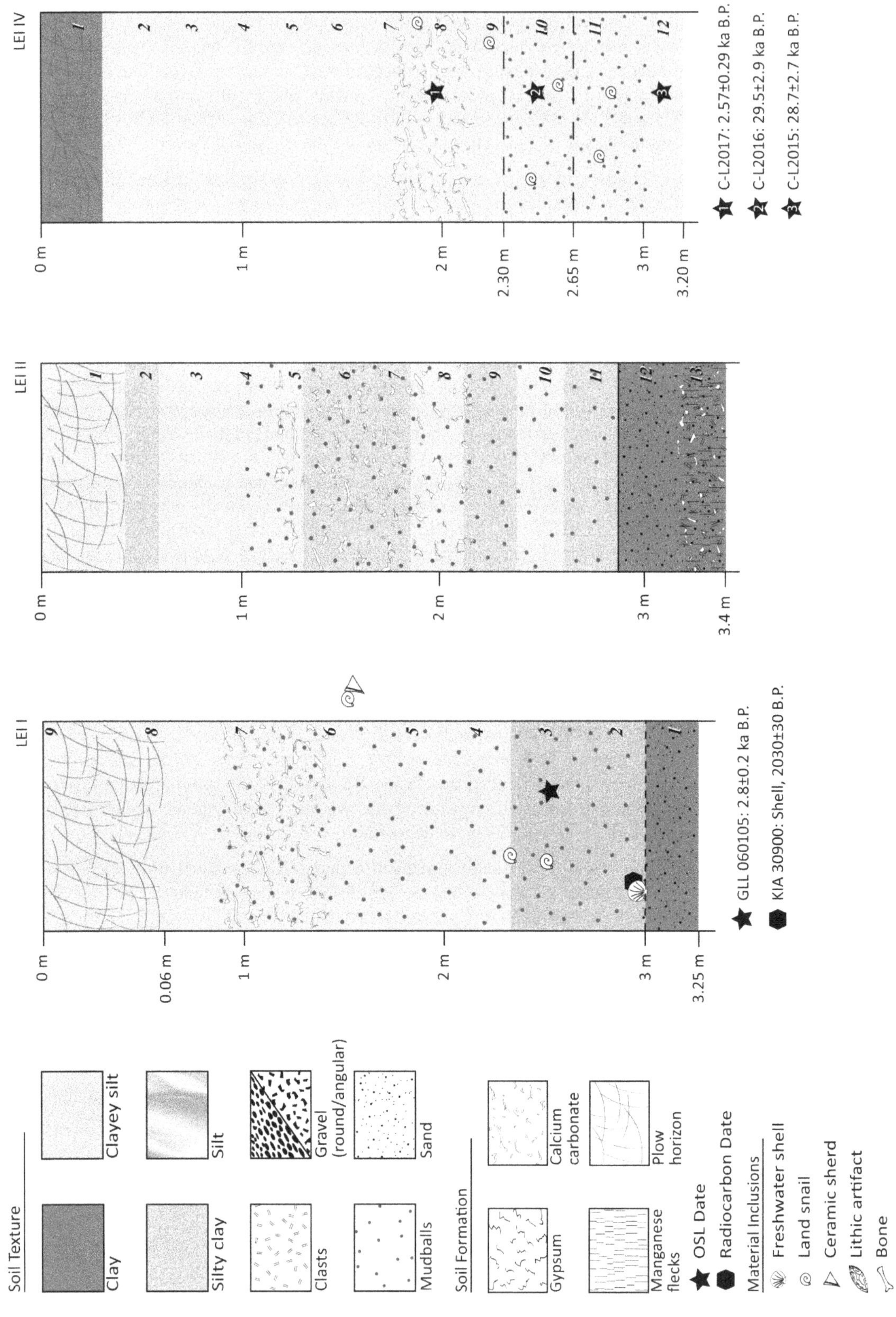

Fig. 2: LEI I, LEI II, and LEI IV sequences along the Wadi Siblah.

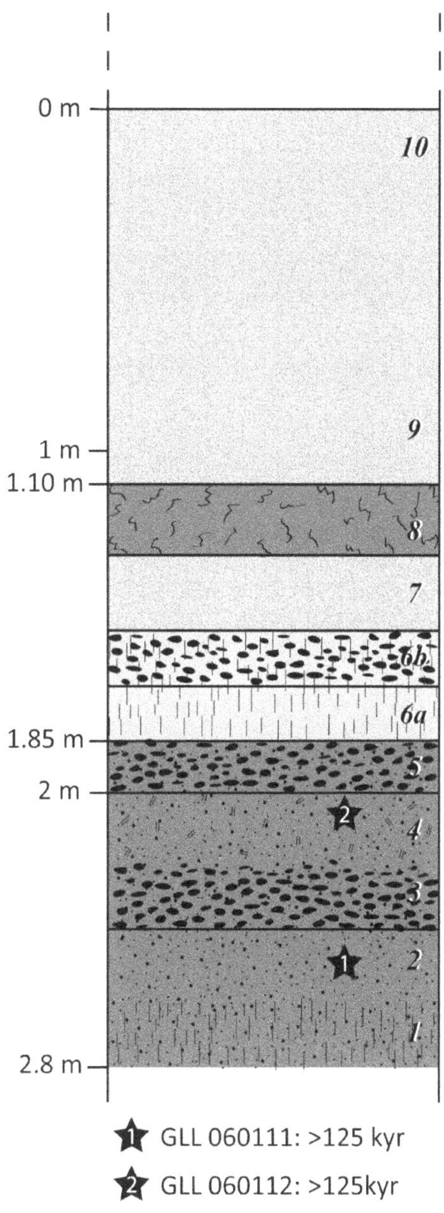

Fig. 3: LEI VII sequence along the Wadi Siblah.

centimetre thick gypsum layer is topped by more than one metre of silty clay. The appearance of this profile, with cementation and intensive manganese staining, suggest it is much older than the other ones. The dates support this idea, in that they are more than 125 kyr BP (Middle Pleistocene).[16]

3.2 Wadi Jarrah

About 3.8 km north of Leilan, profile SOUD was investigated and revealed to be thoroughly comprised of clayey silt (fig. 4). This deposit appears to have accumulated for some time because clear soil formation was visible within the sediments and consists of $CaCO_3$ accumulation to a depth of between 0.9 and 1.6 m. Manganese flecks are visible within the bottom reach of the profile.

About 600 m south of SOUD, the natural 4.1 m deep, exposed profile, SOUD II, was studied (fig. 4). This profile consists of clayey silt that is followed by a graded gravel with (sub-) rounded clasts up to 3.5 cm. The clasts are in a coarse sand matrix that is not imbricated. The gravel is followed by a level of clay and then another gravel deposit that has a clayey silt matrix. Additionally, the gravel contained a lithic

[16] Vandenberghe, this volume chapter 10.

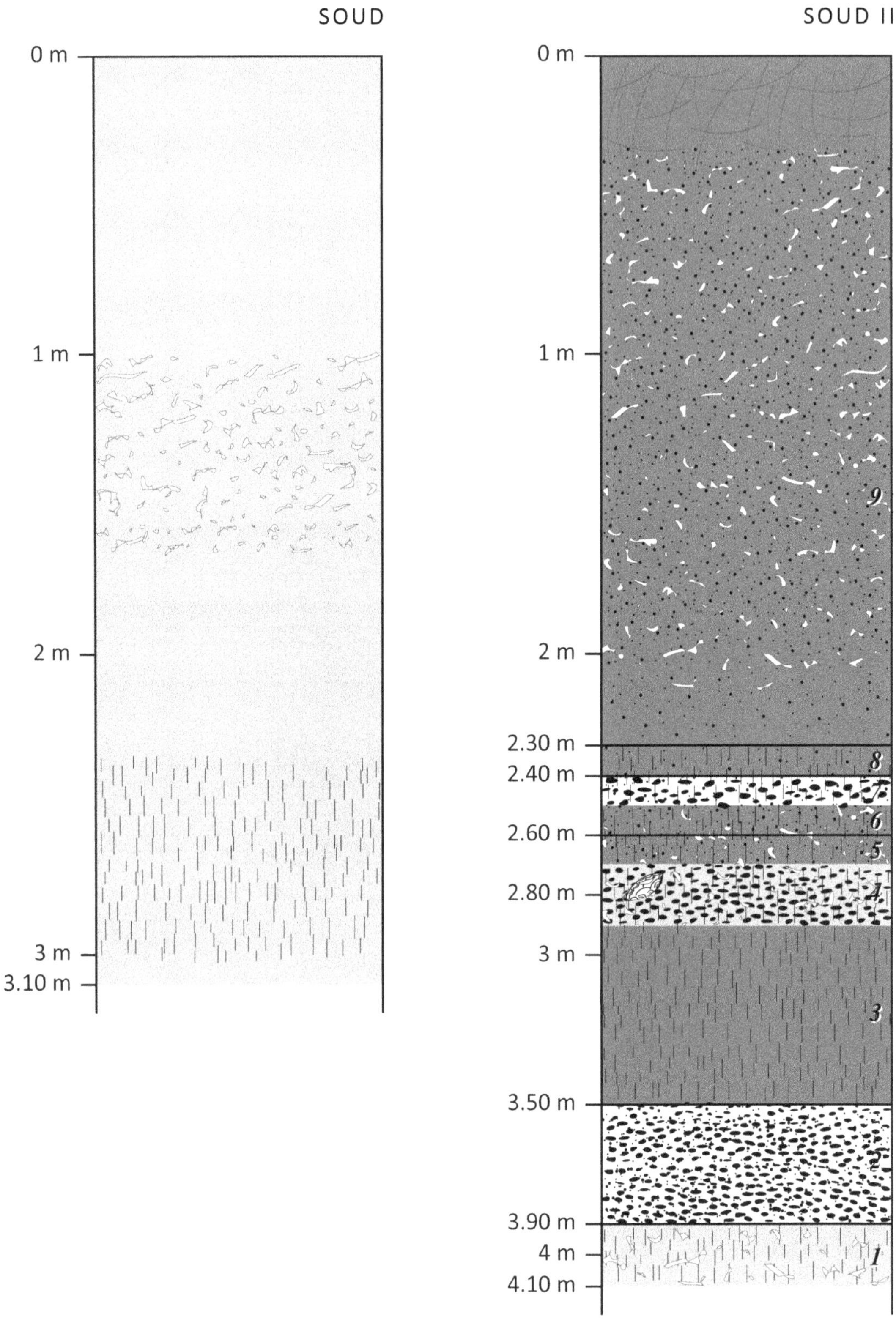

Fig. 4: SOUD and SOUD II sequences along the Wadi Jarrah.

artefact. This last gravel level is then also followed by clay, which in turn is again followed by a thin gravel layer with a fine sand matrix. Above these layers, 2.3 m of sandy clay is deposited. There is considerable evidence for soil formation in the upper part of the profile, consisting of $CaCO_3$ accumulation between 0.3 and 2.2 m. Finally, also noteworthy of mentioning, manganese flecks occur between a depth of 2.3 and 3.5 m.

LEI III (figs. 1 and 5) is located adjacent to a present day track and its 2.7 m deep section was straightened with a bulldozer. This profile contains two thin gravel deposits towards its lower part and many fine-grained overbank deposits. Several OSL samples were obtained but no chronological investigations were as yet undertaken. Strong soil formation with $CaCO_3$ and gypsum flecks between a depth of 1.3 and 2.7 m indicate that some time has passed since the sediments were deposited. The present day plough soil is considerably deep, up to one metre.

LEI V (figs. 1 and 5) is a circa three metre deep bulldozer cut, located about seventy metres west of the present day Wadi Jarrah. At a level of between two and 2.3 m, below the plain, coarse-grained channel deposits in a silty sand matrix were found. The channel is 3.8 m wide and only about twenty centimetres deep. Two OSL dates, one from below this gravel and one at the same level but in the sandy deposits adjacent, provide Late Pleistocene dates (GLL 060109 and GLL 060110).[17] The clasts within the channel are up to five centimetres in size and are rounded to subrounded, which indicates considerable flow velocities. Above this channel lag deposit, sediments are fining upwards to clayey silt. Clayey silt deposits occur between a depth of roughly 1.9 and 0.7 m from the present day surface. They are followed by clayey sand deposits and the present plough zone consists of clayey silt. Besides the 0.3 m rich organic plough horizon, evidence for soil formation is also strong in this profile, particularly between a depth of 0.4 and 1.2 m, where calcium carbonate accumulation occurs.

South of the confluence of the Wadi Jarrah with the Wadi Siblah, just southwest of the lower town of Leilan, a 2.6 m man-made exposure along the present Wadi Jarrah course was studied. This profile, LEI 0, consists mainly of clayey silt with few clasts of fluvial origin (figs. 1 and 5). The level between 0.8 and 0.4 m is the only level that consists of medium sand with no clasts present and is possibly of eolean origin. An undated bone was found at a depth of 0.9 m. Evidence for soil formation within the profile is weak apart from a clear plough horizon. However, sedimentological analysis does show some changes that may relate to incipient soil formation. For example, just below the level with medium-sized sands an incipient soil may have formed that consists of an incipient A-horizon (as indicated by slightly higher organic matter content values and magnetic susceptibility values).

3.3 Wadi Qatrani

Satellite images show a darker area adjacent to the eastern lower town of Leilan (fig. 1). This is related to the floodplain of the Wadi Qatrani and some palaeochannel features are visible within this moister area. LEI VI is a 3.9 m deep bulldozer pit, located about seventy metres west of the present day Wadi Qatrani. It is set within a palaeochannel that is indicated by a darker meandering line on the satellite image (fig. 1). On the surface, in the fields, this channel is hardly visible. At the bottom of the 3.9 m deep pit (fig. 6), very fine gravel deposits within a medium sand matrix were found that represent channel deposits. An OSL sample provided a maximal date of 47 ± 4 kyr BP (GLL 06107).[18] At a depth of 3.4 to 3.2 m below the present day surface another channel lag deposit was found. The channel has a width of 2.7 m and contains small gravels that are fining upwards within a clayey sand matrix. From within these channel deposits an *Unio* shell was retrieved and radiocarbon dated to 5100 ± 30 BP (KIA 30899: 3889 ± 62 calibrated BC). This is a maximal date because we have to take into account an unknown hard water effect.

Additionally, above the channel deposits almost a metre of silty clay overbank deposits were found. Two OSL dates were performed on this unit, the first provided a date of 11.1 ± 0.9 kyr BP (GLL 060108).[19] Considering the shell dated below, this must be an overestimate. The second date, of 4.04 ± 0.4 kyr BP, fits well within the sequence (C-L2018).[20] Gravel in a silty clay matrix was deposited sometime after this date and was followed by 1.5 m of silty clay. Finally, the present day plough soil consists of clayey silt.

[17] Vandenberghe, this volume chapter 10.

[18] Vandenberghe, this volume chapter 10.

[19] Vandenberghe, this volume chapter 10.

[20] Hilgers, this volume chapter 11.

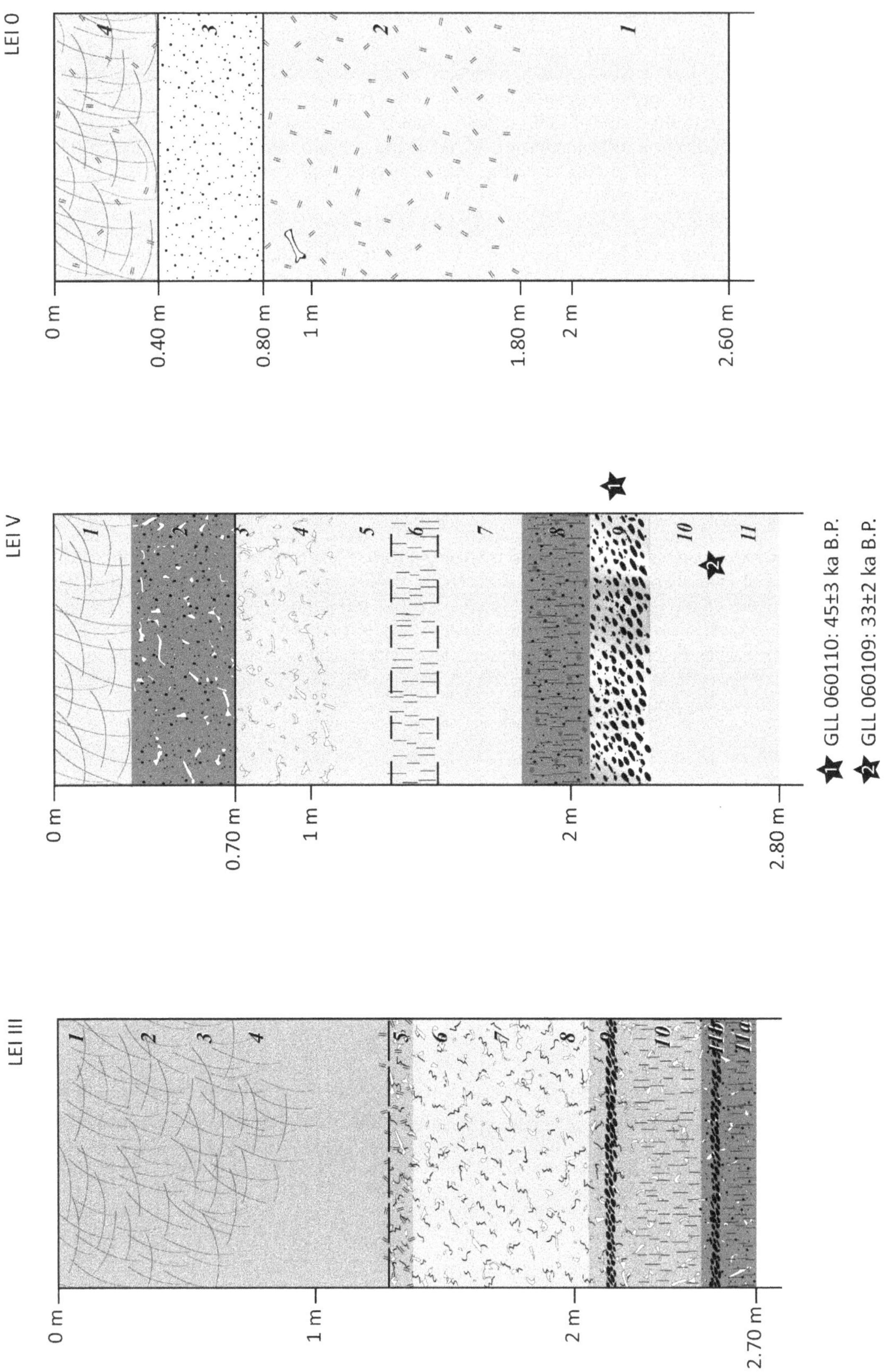

Fig. 5: LEI III, LEI V, and LEI 0 sequences along the Wadi Jarrah.

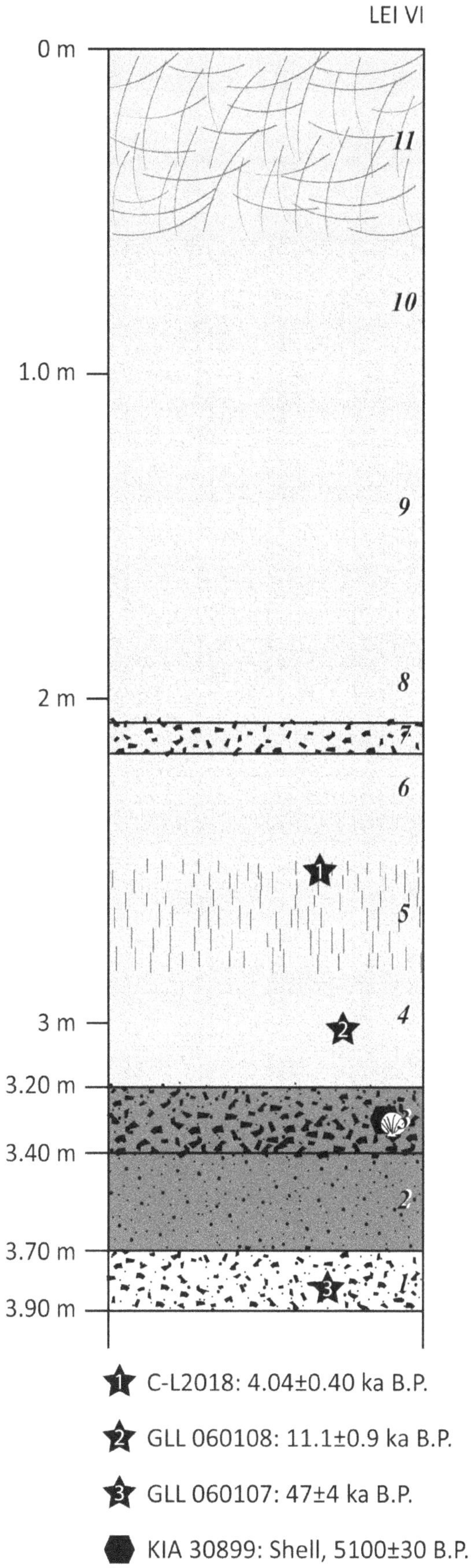

Fig. 6: LEI VI sequence along the Wadi Qatrani.

4. Discussion and conclusion

One of the most interesting features of this research is the fact that canalisation of the Wadi Siblah took place, and that it did so possibly in relation to the construction of the lower town wall, circa 2600 BC. However, it may also be a later feature, at the latest date to circa. 1700 BC, which is the final intensive occupation of Leilan[21]. Weiss *et al.*[22] found evidence of canal works closer to Tell Leilan that they assigned to the Akkadian reorganisation of the site at around 2300 BC. In trench D they found evidence of repeated entrenchments into virgin soil, embankments with large basalt blocks, and masses of waterborn silts and pebbles that were cleared from the channel on the western side of Leilan. The radiocarbon date for the meandering channel that occupied the later stages of this channel also proves it must have been in use before 41 ± 42 calibrated BC. One may question why this feature only exists locally and what the reason was for its existence. Satellite images from Google Earth (fig. 1b) suggest that it was cut to divert the water towards the town because previously, at the height of Leilan, the Wadi Siblah probably joined the Wadi Jaghjagh south of Leilan and was located about 700 m west of the site. The town's intensified need for water may relate to the large settlement expansion that took place circa 2600 BC. The six-fold population increase that occurred would have considerably changed the demands.

The Roman, or later (in case of a strong hard water effect), meandering channel that occupied the canal consisted of clayey sand. Strong sedimentation took place not so long after the channel abandoned this course, as indicated by strong soil formation within these sediments.

To the east of Leilan, in the Wadi Qatrani and presumably dating to the fourth millennium BC, or perhaps to the early third millennium BC, a 2.7 m wide riverbed was found in sequence LEI VI. The early third millennium BC date for the channel could be possible because, as indicated from the HAM sequence, the hard water effect on *Unio* shells in this region may only maximally be about 1000 years.[23] The deposition of about one metre of overbank deposits by 4.04 ± 0.4 kyr BP indicates the continued presence of river activity at that time, but the channel must have been located somewhere else. Moreover, the presence of several channel lag deposits, one above another, indicates an aggrading riverine environment. This may be related partially to intensive agricultural activities that took place within the surroundings. It is interesting that the matrix of the channel deposits changes from medium-sized sand, to clayey sand, to silty clay; it may be related to lower flow velocities over time. This reduction in wadi velocities could be related to climatic drying,[24] however, it may also be an affect of increased water-use by humans.

The rounded and fairly large gravels found within the Wadi Jarrah palaeosediments indicate that this river had the highest palaeovelocity of the three wadis that passed near Leilan. The gravel sheets within the sequences, which are sometimes visible over longer distances, may not actually indicate a single channel but rather a broad gravel sheet that was within a braided river system. This fits the find of Besonen and Cremaschi.[25] Moreover, the presence of a fair amount of fine-grained sediments also suggests the existence of a meandering system at some time, but no certain chronology could be attained. It is possible that the coarse-grained channel deposit found in LEI V is of Pleistocene age; however, it just as likely may be an overestimate.

Section LEI 0 seems to date to the period after the canal was constructed. Of particular interest is this section's possible eolean sand deposit. Could it relate to the sand deposit found on the site, which is dated to the abandonment? Unfortunately, eolean sedimentation is not rare for the region and may also relate to later periods. Tell Brak is one such example from the region where eolean deposits that were laid down over an extended period of time were found.[26]

Bibliography

BESONEN, M. – CREMASCHI, M.

2002 Geomorphological Field Survey Report. http://leilan.yale.edu/works/geo_report/index.html (accessed June 25, 2009).

DECKERS, K. – RIEHL, S.

2007 Fluvial deposits from northeastern Syria: Environmental contexts for archaeological sites, Quaternary Research 67, 337-348.

2008 Tells of the Upper Khabur: their hydrological and agrarian contexts, Paléorient 34.2, 173-189.

---

[21] Risvet – Weiss 2008.

[22] Weiss *et al.* 1993.

[23] Deckers, this volume chapter 7.

[24] Wick *et al.* 2003.

[25] Besonen – Cremaschi 2002.

[26] Wilkinson – Deckers, this volume chapter 9.

MEIJER, T. – PREECE, R.C.
2000 A review of the occurrence of *Corbicula* in the Pleistocene of North-West Europe, Geologie & Mijnbouw 79 (2/3), 241-255.

RISVET, L.
2005 Settlement, Economy, and Society in the Tell Leilan Region, Syria, 3000-1000 BC, Unpublished PhD dissertation University of Cambridge.

RISVET, L. – WEISS, H.
2008 Micro- and macro-contexts of the Tell Leilan tablets, in J. Eidem (ed.), The royal archives from Tell Leilan: Old Babylonian letters and treaties from the lower town palace.

STAUBWASSER, M. – WEISS, H.
2006 Introduction: Holocene climate and cultural evolution in Late Prehistoric-Early Historic West Asia, Quaternary Research 66, 372-387.

WEISS, H. – COURTY, M.A. – WELLERSTROM, W. – GUICHARD, F. – SENIOR, L. – MEADOW, R. – CURROW, A.
1993 The Genesis and Collapse of Third Millennium North Mesopotamian Civilization, Science 291, 995-1088.

WICK, L. – LEMCKE, G. – STURM, M.
2003 Evidence of Lateglacial and Holocene Climatic Change and Human Impact in Eastern Anatolia: High-resolution pollen, charcoal, isotopic and geochemical records from the laminated sediments of Lake Van, Turkey, The Holocene 13.5, 665-675.

7. Riverine development in the Tell Hamidi surroundings

Katleen Deckers

1. Introduction

The Wadi Jaghjagh is the largest tributary of the Upper Khabur and is the wadi considered to be of greatest importance within the region (fig. 1). In this chapter, the fluvial development of the Wadi Jaghjagh in the surroundings of Tell Hamidi is investigated.

Although the Jaghjagh is naturally a permanent stream absent of damming and irrigation, it presently runs dry during the summer south of the town of Qamishli. Flow data of the Jaghjagh published in Kerbe[1] indicates that already in the fifties, from which his earliest flow data derive, there were some years when the flow was interrupted for plus or minus one hundred days a year. This was probably the result of damming in Turkey. The average discharge at Qamishli for over twenty-five years, between 1951 and 1975, was 4.3 m^3/s. Figure 2 shows the annual and inter-annual Jaghjagh flow variations measured at Qamishli for the 1951-1975 period. Kerbe notes that although there was a dam in Turkey, the measurements between 1954 and 1957 are most representative of the natural flow fluctuations of the Jaghjagh. From the sixties onwards, due to increasing water use, the Jaghjagh ran dry in many years for several months at Qamishli's measurement station. This usually took place in the months of July, August, and September. In contrast, maximal flow generally occurred between November and May with peaks up to 43.4 m^3/s, as seen in December 1967.[2]

Photographs from the Gertrude Bell Archive document Wadi Jaghjagh at the Nisibis Bridge in April 1911.[3] From the pictures it can be estimated that the visible channel of the Jaghjagh must have been about nine metres wide. Judging from the shape of the bridge however, there appear to have been three channels at this reach that suggest it had an anastomosing morphology. All of the photographs indicate that the Jaghjagh seems to have contained no riparian vegetation at this location. This is further supported by Poidebard,[4] who also made several images of the Jaghjagh, including aerial photographs. In all of his photographs the wadi contained water but it had no accompanying riparian woody vegetation. As visible from one aerial photograph, from the Barri area (roughly eleven kilometres south of Hamidi), it had a meandering morphology at this reach. Poidebard's images also include a dam on the Jaghjagh with a height of circa ten metres. In contrast to the aforementioned photographs, Dilleman[5] published one of the Jaghjagh adjacent to a gallery forest.

The Jaghjagh is often the only wadi of the Upper Khabur Basin that is mentioned in geographical studies and it has also been referred to in several historical sources. Many archaeological sites are located along its course (fig. 1). Particularly interesting is that the Jaghjagh even played a role in the siege of the city of Nisibis in 350 AD. Although the five sources about the battle differ markedly in detail, they all agree that the Wadi Jaghjagh (at that time named "Mygdonius") was involved and was made to cause a degree of flooding, which led to the collapse of part of the Nisibis city wall.[6] For these reasons, the Wadi Jaghjagh is truly an important river and knowing its fluvial history will shed light on the environmental contexts of many adjacent archaeological sites.

To better understand the Jaghjagh's history, four natural exposures and three borehole profiles were studied just outside the lower town of Tell Hamidi, a well-known excavated archaeological settlement mound with a lower town (figs. 1 and 3).[7] Investigated occupation levels at this archaeological site span from the second millennium BC up until recent times. The location of the examined profiles are visible in Figure 3. A main transect of profiles was studied about 650 m north of the tell (fig. 3), just outside the town wall, and consists of: boreholes HAMB1, HAMB2, and profile HAMIII on the left bank (looking

---

1 Kerbe 1987.

2 Kerbe 1987, 718.

3 http://www.gerty.ncl.ac.uk/

4 Poidebard 1934.

5 Dilleman 1962, Pl. 5.

6 Lightfoot 1988.

7 Wäfler 1990; 2001.

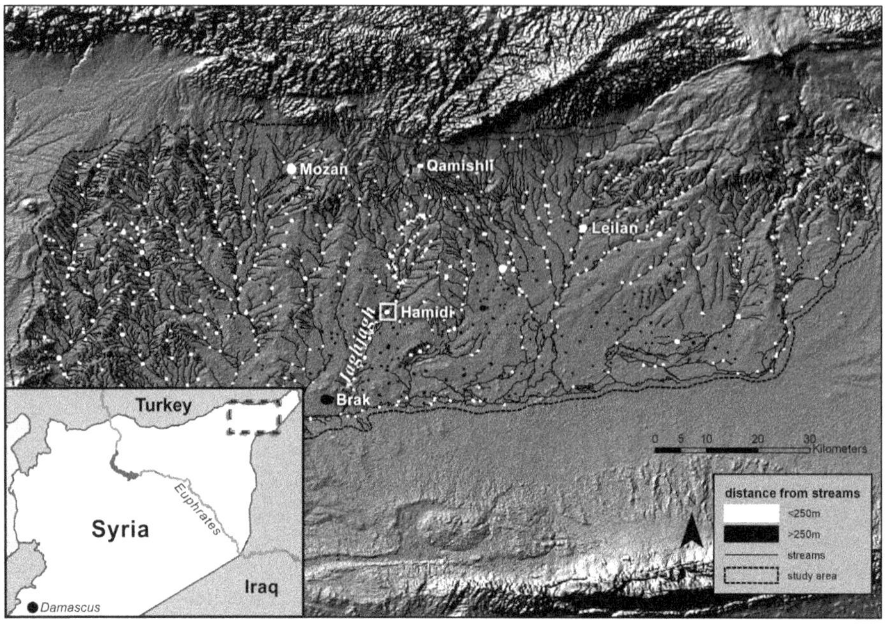

Fig. 1: Map of Upper Khabur with location of Jaghjagh and area under investigation.

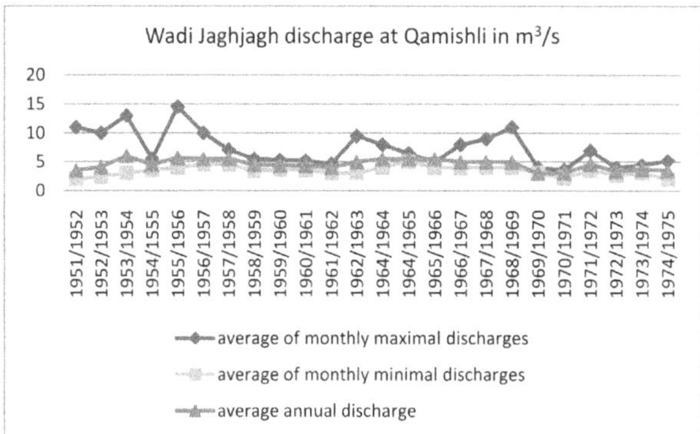

Fig. 2: Jaghjagh discharge measured at Qamishli (data from Kerbe, 1987).

downstream); and borehole HAMB3 and profile HAMIV on the right bank. About 400 m north of this group of profiles (fig. 3), where the track to Hamidi is crossing the Jaghjagh, a natural exposure exists and was designated as HAMVII. Finally, south of Hamidi, another profile – HAMII – was investigated (fig. 3).

2. Methods

The profiles were studied in the field using standard geomorphological and soil science methods.

Special care was taken to retrieve samples to gain chronological insight. For example, pottery sherds were retrieved and typologically investigated, but also TL-screened.[8] Moreover, optically stimulated luminescence (OSL) sediment samples and radiocarbon samples, mainly *in situ* bivalves and organic remains, were also retrieved. The OSL samples from the profile were investigated by Dr. Dimitri Vandenberghe.[9] Additionally, radiocarbon samples were analysed in Kiel by Prof. Piet Grootes and his team.

[8] After the methodology described in Deckers – Vandenberghe 2007; Deckers 2010.

[9] see Vandenberghe, chapter 10 this volume.

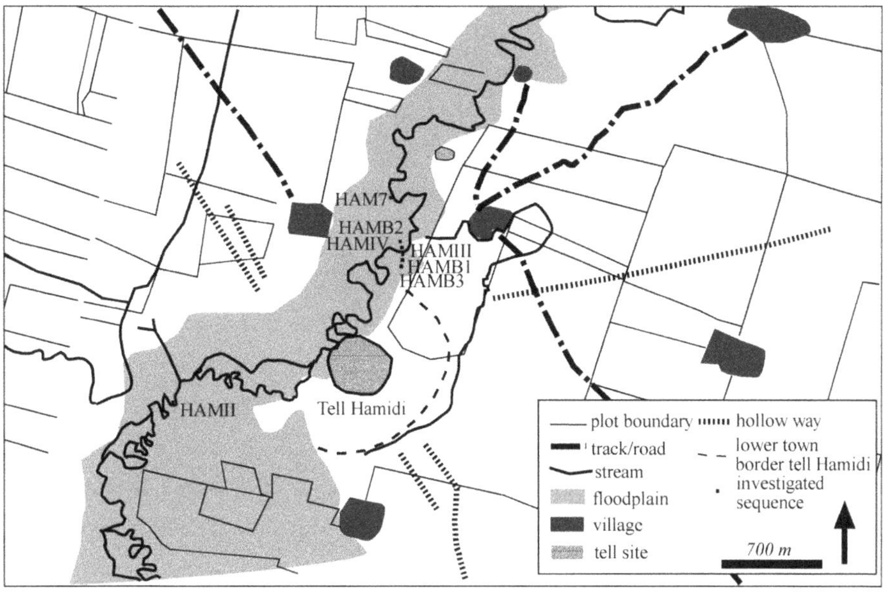

Fig. 3: Detailed map of the Tell Hamidi region according to CORONA satellite image D025-055 1105-1FWD, 5 Nov 1968.

Table 1: Clast measurements from the investigated Jaghjagh profiles with calculated palaeovelocities and estimated discharge.

	Average clast size (mm)	Maximum clast size (mm)	Average velocity (m/s)	Maximum velocity (m/s)	Date	Average discharge (assuming an 18 m width and 2 m depth)
HAMII.1	115	230	1.6	2.1	younger than 1.5 ± 0.6 kyr (TL screened sherd)	77
HAMII.2	62	370	1.2	2.6	younger than 1.5 ± 0.6 kyr (TL screened sherd)	95
HAMII.3	87	150	1.4	1.8	younger than 1.5 ± 0.6 kyr (TL screened sherd)	64
HAMIII.2	132	250	1.7	2.2	mid 4th millennium Cal. BC	80
HAMIII.3	146	1130	1.7	4.3	mid 4th millennium Cal. BC	155
QAMII.6	88	150	1.4	1.8	2325 ± 25 ^{14}C BP (snail) - 5th century BC or later	64
QAMII.4	85	310	1.4	2.4	2380 ± 30 ^{14}C BP (snail) - 5th century BC or later	88
QAMII.2	166	390	1.8	2.7	>2380 ± 30 ^{14}C BP (snail) - 5th century BC or later	97
QAMVIII	178	480	1.9	3.0	younger than 1.1 ± 0,4 kyr (TL screened sherd)	106
QAMVI	149	1270	1.8	4.5	NA	163

At a level of roughly four metres below the plain, in two of the cores and in one profile, an *in situ* organic-rich and water-saturated layer was discovered. Samples from it were then floated to retrieve the botanical remains; with the fruits and seeds examined by Dr. Simone Riehl[10] and the wood (charcoal) remains by the author. Additionally, Dr. Riehl also investigated ostracod remains from several samples.[11]

An idea of the magnitude of river flows can be gained from the application of a mathematical equation, which attempts to relate a given particle to the velocity that would be required to transport the specified particle size. The said equation is applied in this research as:

$$V_b = 0.64\ D^{0.44}$$

The mean velocity at a specific distance from the bed is expressed as 'V_b' and is determined by the particle size in feet per second. 'D' is the particle size in millimetres. A conversion factor of 1.4 is applied to correct V_b to the mean flow velocity if, as seems reasonable to assume in this case, the flow depth is several times the particle diametre.[12] In this study, 50 clasts for each sample from gravel sediments were measured (tab. 1). The discharge of the river was gained by multiplying the velocity with the depth and width of the channel.

---

[10] see Riehl, chapter 8 this volume.

[11] see Riehl, chapter 8 this volume.

[12] Gomez 1987, 354.

Fig. 4a: Illustration of HAMVII; 4b. Photograph of HAMVII.

Table 2: Radiocarbon dating results.

Sample number	Profile and sample	Material	Radiocarbon date	Calibration	$\delta^{13}C$(‰)
KIA 31308	HAMVII.3	bivalve	2455 ± 30 BP	596 ± 120 BC	-9.73 ± 0.10
KIA 31309	HAMIII-below org. stratum	bivalve	5725 ± 40 BP	4584 ± 62 BC	-10.01 ± 0.15
KIA 31663	HAMIV.3 shell from recent deposit	-	1635 ± 25 BP	434 ± 42 AD	-9.79 ± 0.20
KIA24908	HAMB1.13	charred plant fragment	4665 ± 25 BP	3449 ± 51 BC	-25.98 ± 0.06

3. Profiles

In September 2006, profile HAMVII (figs. 3 and 4) was studied in the field as part of the Tell Brak survey with the guidance of Prof. Tony Wilkinson and Prof. Henry Wright. A channel profile filled with darker fine-grained deposits was visible in this profile. This channel had a depth of more than two metres and appeared to be wider than eighteen metres. At the bottom of HAMVII a forty centimetre-thick gravel is observed that contains clasts of up to six centimetres in size. This gravel layer is abruptly followed by more than two metres of fine-grained, dark vertical accretion deposits that fill the majority of the channel. Because the bulk of the sequence consists of these vertical accretion deposits, it is evident that it represents a typical oxbow fill caused by a neck cut-off. Additionally, several *Unio* bivalves were found in the gravel, of which one was dated at 2455 ± 30 BP (KIA 31308) (tab. 2).

Another profile, HAMIII, is exposed at the present day cut bank (figs. 3, 5, 6, and 7) and was investigated in combination with several boreholes to gain increased insight into the lateral fluvial changes.

The HAMIII profile (fig. 6) contains a typical point bar sequence, with gravels and trough-cross bedded gravels and ripple-cross laminated sands, followed by vertical accretion deposits. Within the upper part, soil formation took place that may have homogenised previous laminations. Similar sequences of small gravels followed by about 2.5 m of clayey silt were visible in the boreholes HAMB3 and HAMB1 (fig. 5).

Towards the right side of the investigated exposure of HAMIII (fig. 6), within a sandy layer that presumably represents the wadi bank, many palaeobotanical remains was found. Some of these organic finds are

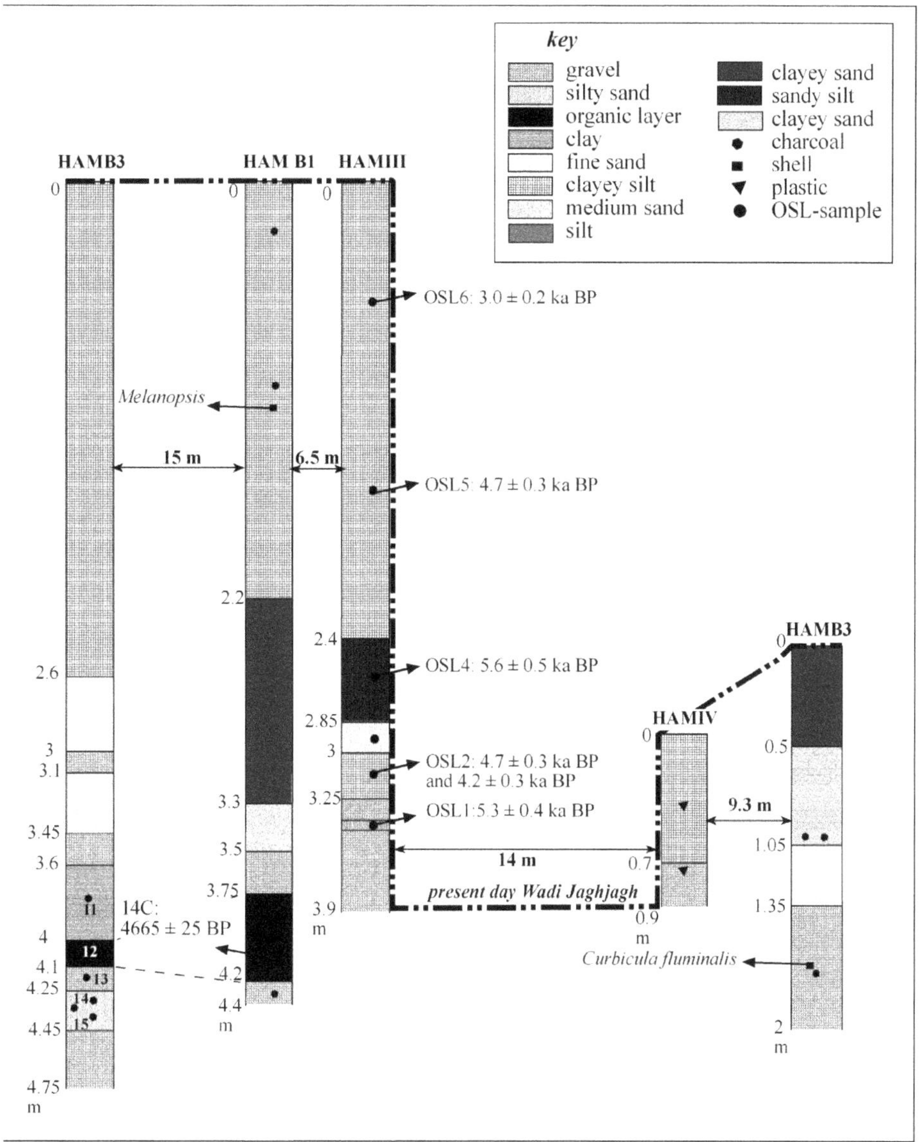

Fig. 5: Hamidi transect with HAMIII, HAMB1, HAMB2, HAMB3, and HAMB4 marked.

preserved uncharred while others are charred. They were clearly laid down by water, since laminations are visible (fig. 8). Several bags of this sediment were floated and seeds, fruits, and ostracods were analysed by S. Riehl,[13] while woody plant remains were identified by the author (tab. 3). Interestingly, this layer contained many *Phragmites* fragments; the largest proportion of it was charred with a smaller proportion left uncharred. Additionally, charred woody remains of *Rubus* were found in this layer, of which also charred fruit remains were discovered.[14] Most of the other carbonised plant seeds represent weeds.[15]

The organic-rich layer of HAMIII correlates to a similar organic-rich layer recovered in boreholes HAMB3 and HAMB1. Like with HAMIII, the botanical and ostracod remains of these boreholes have also been studied in great detail.[16] Of special interest is the *Salix* charcoal piece (tab. 3) and uncarbonised buds that were recovered from this layer of borehole HAMB1. Additionally, *Cyperus longus* remains were also found there. A radiocarbon date of 3449 ± 49 calibrated BC (KIA 24908) exists for this level. The presence of this organic-rich layer four metres below the plain at several locations hints at there being a palaeosurface at this

[13] chapter 8 this volume.

[14] see Riehl, chapter 8 this volume.

[15] see detail in Riehl, chapter 8 this volume.

[16] see Riehl, chapter 8 this volume; tab. 3.

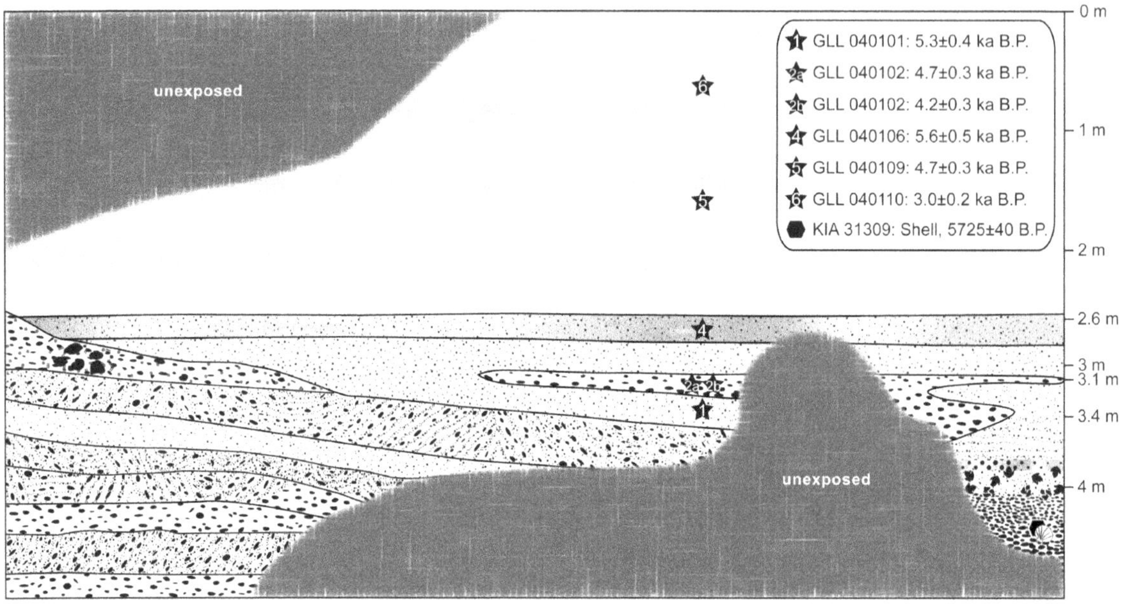

Fig. 6: Detailed profile illustration of HAMIII.

Table 3: List of woody plant remains from the Hamidi transect.

	HAM III	HAM B3.12
Phragmites charred	234	-
Phragmites uncharred	11	-
Rubus charred	16	-
Salix wood charred	-	5

level and implies that it has aggraded for about four metres after the mid fourth millennium BC. Its presence over a larger area also suggests that these remains may have been (partially) burned locally and subsequently flooded. However, it cannot be excluded that they may represent redeposited cultural material, such as dung.

In the gravel of a river channel deposit, just below the organic-rich layer in HAMIII, one of several *Unio* bivalves was radiocarbon dated to 5725 ± 40 BP (KIA 31309) (fig. 6), giving a calibrated date of 4584 ± 62 BC. Although it is possible that these channel sediments are somewhat younger than the calibrated date, they are not necessarily 1000 years older than the organic-rich layer above. The rather *old* date may be due to the "hard water effect" – something that has not been thoroughly investigated for this region and shell species. From the date given, it can be concluded that the hard water effect must be taken into consideration for *Unio* shells in this region, but that it is maximally only up to 1000 years. Prof. Dr. Wilkinson´s notes record two sherds in these gravels of the earlier 4th millennium BC that suggest an earlier 4th millennium BC date or later.

Additional dates from three OSL samples, GLL 040101, GLL 040102, and GLL 040106 (figs. 4 and 6 and tab. 4),[17] have been gleaned thus far from the point bar deposits of HAMIII. Providing dates between 5.6 and 4.2 kyr BP[18], these samples correspond well with the radiocarbon date from the organic layer and indicate that a Jaghjagh meander was located there at that time. A TL-screening date of 4.1 ± 1.5 kyr BP does not contradict the other dates (tab. 4). Measurements of the clasts indicate that on average the stream velocity must have been between 1.4 and 1.7 m/s (tab. 1).

[17] see also Vandenberghe, chapter 10 this volume.

[18] Dr. Markus Fuchs also conducted a preliminary investigation on an additional OSL sample from this profile. Unfortunately, it appears likely that a sample from the Hamidi sequence was swapped with one from the Kurdo profile (see Deckers - Pustovoytov, chapter 4 this volume). This is speculated for the following reasons: the intensive $CaCO_3$ accumulation within the Kurdo 1 profile and the abnormally high-flow velocity indicates a Pleistocene age. For this reason, a date of 5.75 ± 1.94 kyr BP is very unlikely. Additionally, this date fits well with all other dates from the Hamidi sequence.

Fig. 7a. Location of profile HAMIII and borehole HAMB1 along the Jaghjagh. 7b. Overview of exposure at HAMIII. 7c. Detail of trough cross-stratified gravels and rip-up clasts.

Fig. 8: Detail of organic layer from HAMIII.

Two OSL dates on the vertical accretion deposits, samples GLL-040109 and GLL-040110[19], indicate that vertical accretion took place in the earlier half of the third millennium BC and at about 1000 BC. A sherd from these accretion sediments found at a depth of about 1.8 m, typologically assigned to probably the end of the Early Bronze Age/start of the Middle Bronze Age (ca. 2300-1700 BC) by Prof. Dr. T. Wilkinson,[20] does not contradict this chronological scheme. No distinct palaeosol was found between the sediments dated by the samples GLL-040109 and GLL-040110.[21] This suggests that longer periods of soil formation could not take place because of regular flooding. However, the sediments were slightly darker throughout the profile and no laminations were visible, which suggests that between phases of flooding some plant growth and initial Ah formation took place.

A third investigated profile is HAMII, which is located south of Tell Hamidi and was exposed by the present day Jaghjagh to a depth of 2.8 m (see fig. 3 and 9). At its bottom are two clast-supported gravel layers that were found in a sandy matrix and are separated from one another by clayey sand. The lowermost gravel was very poorly sorted with some clasts measuring up to four centimetres, while most of them were less than two centimetres. The upper gravels are somewhat better sorted but still meagrely so. Both gravel layers contained sherds that were chronologically assigned by TL-screening to 530 ± 589 AD and 883 ± 392 AD (tab. 4). Above the gravel, laminated silt was present and was followed by sand that also contained a sherd. Finally, the upper 1.6 m consists of silty clay deposits in which initial soil formation took place.

4. Discussion

The profiles near Hamidi contain some indications concerning the Jaghjagh morphology at this reach. Firstly, the Hamidi transect with the exposed point bar sequence in HAMIII indicates that it was not a braided system in the mid fourth millennium BC, but that its channel was meandering. An anastomosing morphology, however, cannot be ruled out because in all of the profiles (HAMIII, HAM B1, and HAM B3), spread over a distance of 21.5 m, gravels were found. Nevertheless, the HAMVII sequence, representing oxbow deposits clearly post-dating the seventh century BC, is indicative of a meandering system.

According to the velocity calculations, velocities based on average clast size were between 1.2 and 1.9 m/s. In order to compare ancient with recent wadi discharge data, the depth and width of the channel must be established. Unfortunately, these measurements only exist for the HAMVII sequence, which is nearly eighteen metres wide and roughly two metres deep. In spite of this data, it is unclear whether these values

[19] see also Vandenberghe, chapter 10 this volume.

[20] Personal communication T. Wilkinson, September 2006.

[21] Vandenberghe, chapter 10 this volume.

Table 4: TL-screening results.

	Subsam. 1 (Gy)	Subsam. 2 (Gy)	Subsam. 3 (Gy)	Subsam. 4 (Gy)	Average (Gy)	Average (BP*)	Max. dev. from average (years)	Standard error	Systematic error	Total error (years)
HAMII.1	5.098	4.492	3.841	4.25	4.420	1473	226	15%	25%	589
HAMII.3	2.86	3.369	3.528	3.683	3.360	1120	108	10%	25%	392
HAMIII.2	12.795	10.946	12.388	13.258	12.347	4114	467	11%	25%	1481

* Refers to year of measurement: 2003.

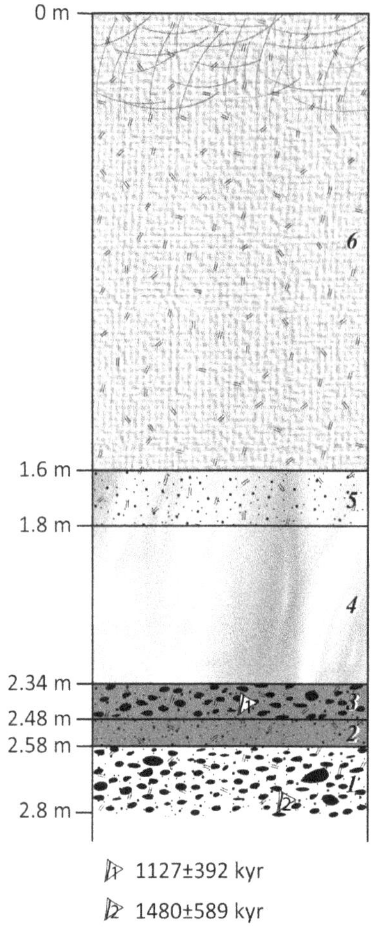

Fig. 9: Profile illustration of HAMII.

can be projected to the mid fourth millennium BC. If we can accept the values from HAMVII to be valid for HAMIII and HAMII, then palaeodischarges of up to 155 m^3/s and 95 m^3/s can respectively be calculated for HAMIII and HAMII. These values are comparatively high since, as mentioned by Kerbe,[22] the maximal debit for the Jaghjagh in recent times was only 43.4 m^3/s. In any case, it is possible that the discharge values are realistic for the fourth millennium BC, considering the fact that point bars are built vertically during bank full and high stage events and that all the discharge values in Kerbe were affected by damming and intensive anthropogenic use. However, it should also be noted that it is probable that the HAMVII profile does not actually represent an orthogonal profile to the main flow; in this way the width of the channel may be somewhat overestimated or a projection of the channel post-dating the seventh century BC cannot be projected back to the past. Overall, though, we can assume that the discharges were larger than the velocities mentioned, since it seems reasonable to assume that the channel was more than one metre wide and also had a depth of more than one metre.

Recent anthracological research from Tell Mozan indicates that open oak park woodland was present within this area between circa 2500 and 2000 BC.[23] This is contrary to the tree-less landscape around Mozan that is seen today and that was referenced to in, for example, the Gertrude Bell Archive's photos from the early twentieth century. Although little is known of the earlier vegetation within this area,[24] deciduous oak was probably present during the mid fourth to mid third millennium BC, as is also suggested for present day potential vegetation reconstructions.[25] A denser tree cover caused increasing rainfall interception and inhibited runoff, and may have resulted in steadier perennial flow and raised groundwater tables. Further evidence from northwestern Iraq confirms that raised water

[22] Kerbe 1987.

[23] Deckers – Riehl 2007; Deckers – Pessin, chapter 3 this volume.

[24] see Deckers – Pessin, chapter 3 this volume.

[25] Moore *et al.* 2000.

tables occurred between 6000 and 3000 BC.[26] Wilkinson and Tucker[27] found indications that during the prehistoric period water pits up to 4.5 m in depth were used and that the water table must have been relatively high as indicated by the three to four metres of ground surface mottling. Since the late Uruk period, water levels may have dropped some 115 cm. This may be due to climatic change or because of the removal of vegetation or, alternatively, due to increased abstraction of the groundwater.

Important information was gained concerning the riverine vegetation from the organic-rich layer found within HAMIII, HAMB1, and HAMB3, and dated to the mid fourth millennium BC. It is not entirely clear what this partially burned, organic-rich layer signifies, but its presence at several locations suggests that it most likely represents the partial burning of local vegetation that was subsequently flooded. Several ethnographic studies report native American peoples burning wetland vegetation to improve travel, hunting success, and food availability. Furthermore, native tribes of California also burned reed stands to recycle nutrients, activate rhizomes, and reduce insect pests. In southeastern Louisiana, trappers burned marshes to improve trap accessibility and encourage growth of preferred common muskrat foods, such as common reed. Fires typically burned when soils were wet and caused only minimal damage to marsh vegetation.[28] The Jaghjagh riverine vegetation consisted of Populetea and was accompanied by swamps with *Phragmites* and *Cyperus*.

The presence of a palaeosurface at a level of four metres below the present day floodplain also implies that a depression within the floodplain or the floodplain aggraded by over four metres since the mid fourth millennium BC. Human activities may have contributed significantly to aggradation. The plain was intensively inhabited during the middle of the third millennium BC and also in the early second millennium BC,[29] and the delivery of fine sediments in the shape of plough-wash could have been considerable. This matches agreeably with the sedimentation observed in the Tell Mozan depression.[30]

5. Conclusion

The Jaghjagh fluvial sediments in the Hamidi neighbourhood document a considerable palaeodischarge in the mid fourth millennium BC, the fifth century BC or somewhat later, and in the fifth century AD or somewhat later. The palaeochannel was meandering probably after the seventh century BC, although it may have been anastomosing in the mid fourth millennium BC at this reach. However, more data is needed to support the latter conclusion. There is also evidence for aggradation from the third until the second millennium BC, which may relate to human impact on the landscape. Finally, the Jaghjagh was originally accompanied by a riverine gallery forest and *Phragmites* and *Cyperus* swamps. Today, these are lacking and were already scarce in the early twentieth century AD.

Bibliography

DECKERS, K.
2010 Archaeological sites of the Upper Khabur Basin in their fluvial geomorphological context, in: P. Pfaelzner (ed.), Ausgrabungen 1998-2001 in der Zentralen Oberstadt von Tall Mozan/Urkeš III: Die naturwissenschaftliche Untersuchungen, 365-381.

DECKERS, K. – RIEHL, S.
2007 An evaluation of botanical assemblages from the 3rd to 2nd millennium BC in northeastern Syria, Varia Anatolica XIX, 481-502.

DECKERS, K. – VANDENBERGHE, D.
2007 Confronting luminescence with radiocarbon dates for fluvial deposits in the Upper Khabur basin of northeastern Syria, in: B. Ooghe (ed.), Proceedings of the Broadening Horizons conference, 50-74.

DILLEMANN, L.
1962 Haute Mésopotamie orientale et pays adjacents, IFAB Bibliothèque Archéologique et historique 72.

Gertrude Bell Archive Website, Newcastle University Library, date accessed: November 2009, <http://www.gerty.ncl.ac.uk/>.

[26] Wilkinson – Tucker 1995.

[27] 1995.

[28] Gucker 2008.

[29] Lyonnet 1998; Staubwasser – Weiss 2006.

[30] Deckers – Pustovoytov, chapter 4 this volume.

GOMEZ, B.
1987 The Alluvial Terrace and Fills of the Lower Vasilikos Valley in the Vicinity of Kalavasos, Cyprus, Transactions of the Institute of British Geographers, 335-359.

GUCKER, C.L.
2008 Phragmites australis, Fire Effects Information System, U.S. Department of Agriculture, Forest Service, Rocky Mountain Research Station, Fire Sciences Laboratory, date accessed: November 2009, <http://www.fs.fed.us/database/feis/>.

KERBE, J.
1987 Climat, hydrologie et aménagements hydro-agricoles de Syrie, unpublished PhD dissertation.

LIGHTFOOT, C.
1988 Facts and fiction: the third siege of Nisibis (A.D. 350), Historia 37, 105-125.

LYONNET, B.
1998 La prospection archéologique de la partie occidentale du Haut-Khabur (Syrie du nord-est): Méthodes, résultats et questions autour de l´occupation aux IIIe et Iie millénaires av. n. E., in: J.M. Durand (ed.), Mari et les hourites, Amurru 1, 363-376.

MOORE, A.M.T. – HILLMAN, G.C. – LEGGE, A.J.
2000 Village on the Euphrates.

POIDEBARD, A.
1934 La trace de Rome dans le désert de Syrie. Le limes de Trajan à la conquête arabe. Recherches aériennes (1925-1932), Bibliothèque Archéologique et historique, du Service des Antiquités de Syrie XVIII.

STAUBWASSER, M. – WEISS, H.
2006 Introduction: Holocene climate and cultural evolution in Late Prehistoric-Early Historic West Asia, Quaternary Research 66, 372-387.

WÄFLER, M.
1990 The Excavations at Tell Hamidi, in: S. Eichler – M. Wäfler (eds.), Tall al-Hamidiya 2. Orbis Biblicus et Orientalis 6, 219-228.
2001 Tall al-Hamidiya 4, Vorbericht 1988-2001, Orbis Biblicus et Orientalis 21.

WILKINSON, T.J. – TUCKER, D.J.
1995 Settlement development in the North Jazira, Iraq. A Study of the Archaeological Landscape.

8. Ostracods and seed remains from the Hamidi sequence

Simone Riehl

1. Introduction

The acquisition of ecological data is a prerequisite for reconstructing environmental dynamics in the past and enables the understanding of landscape formation processes.[1] Palaeoecological research in wetland environments has a long tradition in archaeology and involves, aside from palynology, the evaluation of a broad range of different biological remains, which are today appreciated for their ecological indicator quality.[2] Due to the specific preservation requirements of such remains, their evaluation in Near Eastern archives and archaeological contexts has so far been limited.

Plant remains in semi-arid environments are usually preserved only once they have been carbonised; for this reason, they are abundant in archaeological sites where human-induced fires may have occurred frequently. In most cases they represent crops and weeds from ancient fields. Uncarbonised plant remains are preserved only under absence of microbial activity due to hyper-arid conditions, as for example in royal Egyptian tombs or under exclusion from oxygen, such as can be found in submerged cultural layers of lakeside settlements.

Only recently have *Ostracoda*, representatives of the Crustaceae class populating different types of water bodies, been analysed from Near Eastern lakes to reconstruct lake and climate history.[3] *Ostracoda*, however, are also frequently found in archaeobotanical samples if mesh sizes smaller than 1 mm are used during sample processing.

Near Eastern alluvial plains may potentially exhibit preservation conditions that enable preservation of uncharred plant remains, for example, under fast and quasi-hermetic sedimentation. Therefore, river dynamics and the associated sedimentation process may preserve changing vegetation cover as well as environmental information from organisms with ecological indicator qualities, such as ostracods, throughout time.

This was the case in an organic layer recovered in the Wadi Jaghjagh in northeastern Syria at about 4 m below the surface.[4] This layer provided a radiocarbon date of 3449±49 calibrated BC.

The Jaghjagh, originally a perennial stream due to constant flow from the karstic system in the northern Syrian-Turkish border area,[5] presently runs dry during the summer season because of increased water withdrawal for irrigation and other purposes. The area is almost devoid of natural vegetation today and is primarily used for agriculture (fig. 1).

The following contribution aims to demonstrate how ecological information can be extracted from plant remains and ostracods in Near Eastern alluvial sediments to reconstruct ancient landscapes and to contribute to the understanding of landscape dynamics in an area considered today as incapable of supporting sustainable development.

2. Methods

The samples were taken from two profiles, HAM B3 and HAM III, in a sequence just outside the northern lower town wall of Tell Hamidi. They were removed from a 20-40 cm thick organic layer that occurred at a depth of 4 m below the plain. The HAM B3 and HAM III profiles are about 21.5 m apart from each other and represent the same edge of the wadi bank.[6] In the bore core of HAM B3 the organic layer (HAM B3.12) was sampled as well as an underlying transitional layer of gravel and silty sand (HAM B3.14).

---

1 Lowe – Walker 1997.

2 Evans – O'Connor 2001; and representative for the materials used here: Bertsch 1932; Griffiths – Holmes 2000.

3 Pachur – Hoelzmann 1991; Griffiths *et al.* 2001.

4 see Deckers, this volume chapter 7; Deckers – Riehl 2007.

5 Wirth 1971.

6 for further details and profile description see Deckers – Riehl 2007.

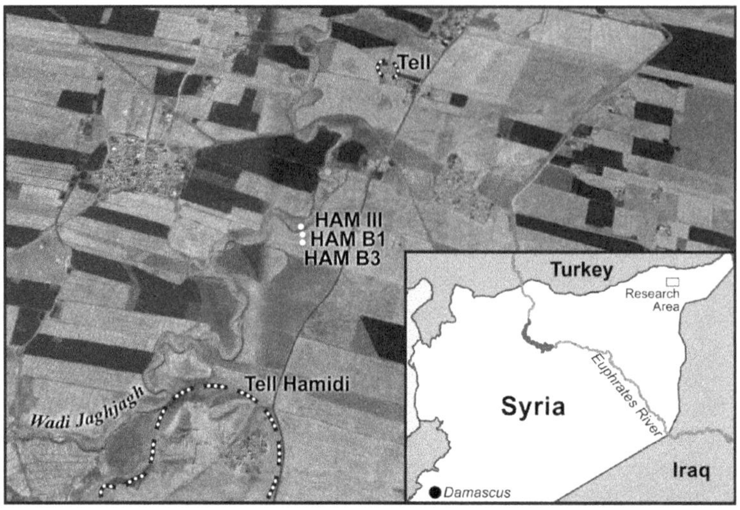

Fig. 1: The Wadi Jaghjagh with location of the sampled profile sequence (HAM B3 and HAM III).

The organic layer was water-saturated and judging from the very low decomposition of this layer it was hoped that uncarbonised plant remains would also be found. Such plant remains would represent another category of interpretation, as they usually correspond to accumulations of wild plant remains from their natural habitats (*biocenosis*), in contrast to accumulations of plant remains under human influence or from human activity (*thanatocenosis*).

After wet-sieving of the samples, plant remains and ostracods were selected from the anorganic fraction under 10-30x magnification using a stereomicroscope.

The sorted plant remains and ostracods were then analysed with the help of standard identification catalogues[7] and in the case of the plant remains, with the additional use of a modern comparative collection.

Ecological assessment of the plant remains was conducted with the information available in floristic literature for the area[8] and in autecological reference works.[9] Ecological data for the ostracods were mainly taken from Meisch.[10]

Archaeobotanical data for the economic background analysis were extracted from the "Archaeobotanical Database of Eastern Mediterranean and Near Eastern Sites".[11]

3. Results and discussion

3.1 The origin of the plant and animal remains

The two different types of preservation of plant remains, carbonised and uncarbonised, indicate the presence of at least two different ways of genesis of the find contexts. While carbonised plant remains mainly represent crops and typical weeds, the uncarbonised remains largely represent wild species from different kinds of habitats (see tab. 1).

The carbonised plant remains arrived at their location of final deposition through water transport, perhaps from settlement sites in close proximity. Because only an individual organic fragment was dated, there may be a chance that some of the charred plant remains are older than 3449±49 calibrated BC; for example, if a sediment sequence from the tell would have been eroded.

The plant remains are very well preserved with minimal signs of corrosion, which reduces the likelihood of their having been transported over a long distance. Considering the flow direction, it is most likely that the carbonised remains were washed ashore from one of the numerous tell sites to the north

[7] listed in van Morkhoven 1963; Hartmann 1964; 1989; Kempf 1973; Freels 1980; Jensen 1998; Martens – Ortal 1999; Griffiths – Holmes 2000; Meisch 2000.

[8] Post 1932; Mouterde 1966.

[9] Ellenberg *et al.* 1991.

[10] Meisch 2000.

[11] Riehl – Kümmel 2005.

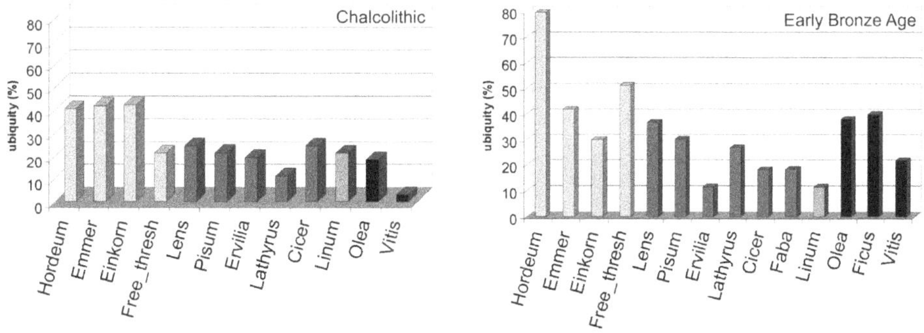

Fig. 2: Frequency of crop species in ninety-seven Chalcolithic and one hundred EBA sites in the Near East (source: extended version of Riehl – Kümmel 2005).

of the profile sequence. Because the lower town wall of Tell Hamidi almost reaches the profile sequence from the south, it is likely that the carbonised plant remains derive from this settlement.

Although most of the carbonised wild plant seeds represent weeds, there are also some other carbonised species that are untypical weeds, which rather indicate the presence of open vegetation, swamps (*Cyperus longus*), and riverbank environments (*Rubus* spp.), or at least of moist soils. The fact that these seeds are carbonised may be interpreted in two ways related to their origin. They may, in fact, have arrived with the carbonised crop remains as part of the cultural layer, and in this case may represent dung remains. Or they could be autochthonous and may have been deposited after burning of vegetation nearby.

3.2 The economic background during layer formation

Aside from the ecological information gained from uncharred plant remains, the carbonised remnants reflect ancient plant production patterns at a settlement site in the neighbourhood of the deposited layer. Furthermore, the latter derive from one of the surrounding settlements and could represent Chalcolithic crop production.

Ubiquities of Chalcolithic crop plant remains from ninety-seven archaeological sites show an interesting phenomenon in comparison with the Early Bronze Age (EBA) (data from one hundred sites). While the crop spectrum is relatively balanced during the Chalcolithic, during the EBA there are clear peaks of frequency for barley, lentil, and the fruit trees. These crop species dominate other crops in frequency and therefore can be interpreted as indicating an economic focus on higher production of these crops. This phenomenon relates to the "shift from emmer wheat to barley" with the transition from the Chalcolithic to the EBA, as recognised by McCorriston in the middle Khabur area and understood to indicate an emerging specialisation in pastoral production.[12] It has also been described by Hald and Charles[13] at Tell Brak as a reduction of the range of crops. While these interpretations seem likely if only emmer wheat and barley are considered, the most likely interpretation is, however, the development of monoculture during the EBA, considering the full crop assemblage in a large number of archaeological sites.

Archaeobotanical remains, in comparison with other palaeoclimate proxies, isotope data from cereal remains, and pedogenic carbonate remains, have been interpreted to indicate regional differences in climate dynamics.[14] Through focussing on the northeastern Mesopotamian area and in considering distribution maps of proportions of crop species intolerant of drought stress during the Chalcolithic period and the EBA, it becomes visible that during the EBA the drought-intolerant crops, einkorn, flax, and garden pea, are either more frequently cultivated along the Khabur or occur further south than during the Chalcolithic.[15] This has been understood as signifying better moisture availability during the EBA than during the Chalcolithic period. The economic answer may have been increased surplus production and the development of monoculture.

Particularly problematic is the fact that to date there is no published $\delta^{13}C$ data from cereals of Chalcolithic sites to allow for reconstruction of drought stress on crops during this period. Unpublished data from Tell Shiukh Fawqani, dating to about 3300 BC, delivered mean $\delta^{13}C$ values of 15.8‰. This generally equals to

[12] McCorriston – Weisberg 2002.

[13] Hald – Charles 2008.

[14] Riehl *et al.* 2008.

[15] Riehl *et al.* 2008, fig. 3.

Table 1: Ecological information from Zohary 1973; Ellenberg et al. 1991; Meisch 2000; Altinsacli – Griffiths 2002.

Family	Taxon	HAM B3.12	HAM B3.14	HAM III	Ecology & life form
	Ostracods				
Candonidae	*Candona* cf. *neglecta* group	2	3	24	Tolerates a wide range of aquatic habitats but prefers cold waters
Cyprididae	Cyprididae (cf. *Bakunella*)	1			-
Cyprididae	*Prionocypris zenkeri*			61	Found in waters connected to springs; avoids deep as well as stagnant waters and prefers cool waters with rich aquatic vegetation
Ilyocyprididae	*Ilyocypris* cf. *bradyi*			45	Prefers springs, slow waters flowing from springs, and ponds by springs
Ilyocyprididae	*Ilyocypris* cf. *inermis*	6	1		Mainly in the bottom mud of springs, ponds fed by springs, and waters flowing from springs; prefers cold temperatures
Ilyocyprididae	*Ilyocypris decipiens*	1			Found in ponds, the littoral zone of lakes, slowly flowing brooks and rivers, temporary pools and fountains
	Carbonised plant remains				
Moraceae	*Ficus carica*			1	Crop
Poaceae (cereals)	Cerealia			1	Crop
Poaceae (cereals)	Cerealia, culm node			1	Crop
Poaceae (cereals)	*Hordeum vulgare*			8	Crop
Poaceae (cereals)	*Triticum aestivum/durum*, rachis			1	Crop
Poaceae (cereals)	*Triticum* cf. *dicoccum*			1	Crop
Poaceae (cereals)	*Triticum diccocum/monococcum*, glume base fragment	1			Crop
Poaceae (cereals)	*Triticum dicoccum*, glume base	1	2	20	Crop
Vitaceae	cf. *Vitis* sp. (fragment)		1		Crop; occasionally in Populetea associations
Fabaceae	*Lathyrus* sp.			1	Crop or crop weed
Fabaceae	*Lens* sp.			2	Crop or crop weed
Poaceae	*Hordeum* sp.			1	Crop or crop weed
Asteraceae	*Centaurea* sp.	1			Crop weed
Chenopodiaceae	*Chenopodium* sp.			1	Crop weed
Poaceae	*Aegilops* sp., glume base			1	Crop weed
Poaceae	*Bromus* sp. (long caryopsis)	1			Crop weed
Poaceae	Poaceae, large			6	Crop weed
Ranunculaceae	*Adonis* sp.			2	Crop weed; prefers dry soils
Ranunculaceae	*Ranunculus arvensis*			1	Crop weed; prefers dry to medium moist soils, enriched in nitrogen
Rubiaceae	*Galium* cf. *Aparine*	1		1	Crop weed; nitrogen indicator
Rubiaceae	*Galium* sp.	1	1	60	Crop weed

Family	Taxon	HAM B3.12	HAM B3.14	HAM III	Ecology & life form
	Carbonised plant remains cont.				
Scrophulariaceae	*Veronica* sp.			2	Crop weed; prefers medium moist soils, enriched in nitrogen
Valerianaceae	*Valerianella dentata*			1	Crop weed; prefers dry to medium moist soils, moderately enriched in nitrogen
Cistaceae	*Helianthemum* sp.			4	Open vegetation
Cyperaceae	*Cyperus longus*	1			Hydrophyt living in Phragmites-swamps, river shore, muddy river shore, sea coast, ditches
Liliaceae	cf. *Muscari/Ornithogalum*			1	Open, overgrazed vegetation
Rosaceae	*Rubus* sp.			3	Probably close to river environments; prefers medium moist to moist soils, moderately enriched in nitrogen
Rosaceae	*Rubus* sp., thorns			5	Probably close to river environments
Asteraceae	Asteraceae, endosperm			1	-
Cyperaceae	Cyperaceae			6	-
Fabaceae	Fabaceae			15	-
Polygonaceae	*Rumex* sp.			9	-
	indet. plant remains			8	
	Uncarbonised plant remains				
Fabaceae (crop pulses)	*Cicer arietinum*	1			Crop
Moraceae	*Ficus carica*			1	Crop
Poaceae (cereals)	*Triticum dicoccum*, glume base			1	Crop
Asteraceae	*Sonchus asper*			1	Crop weed; prefers medium moist to moist soils, enriched in nitrogen
Brassicaceae	*Neslia paniculata*, capsule			2	Crop weed; prefers dry to medium moist soils
Caryophyllaceae	*Silene* sp.			1	Crop weed
Chenopodiaceae	*Chenopodium* sp.			1	Crop weed
Fumariaceae	*Fumaria officinalis*	1	1		Crop weed; prefers medium moist soils, enriched in nitrogen
Primulaceae	*Anagallis* sp.			6	Crop weed; prefers medium moist soils
Ranunculaceae	*Ranunculus arvensis*			1	Crop weed; prefers dry to medium moist soils, enriched in nitrogen
Valerianaceae	*Valerianella carinata*			5	Crop weed; prefers dry to medium moist soils
Valerianaceae	*Valerianella dentata*			9	Crop weed; prefers dry to medium moist soils, moderately enriched in nitrogen
Portulacaceae	*Portulaca olearacea*	1		3	Crop weed; desert places, waste places near sea-level; prefers dry to medium moist soils, enriched in nitrogen
Vitaceae	*Vitis* sp.			2	Crop; occasionally in Populetea associations
Boraginaceae	*Heliotropium europaeum*	1	1	13	Disturbed habitats, ruderal, and in fields; prefers dry to medium moist soils
Boraginaceae	*Lithospermum tenuiflorum*			4	Ruderal
Resedaceae	*Reseda* sp.			3	Open vegetation

Table 1: (Continued).

Family	Taxon	HAM B3.12	HAM B3.14	HAM III	Ecology & life form
	Uncarbonised plant remains cont.				
Rosaceae	*Rubus fruticosus* agg.	4	3	5	Forests to open shrubs, and waste places
Verbenaceae	*Verbena officinalis*	1		2	Ruderal, disturbed places, rocky slopes, dry riverbeds, embankments, walls, sand dunes, wood, bushes; on medium moist soils, enriched in nitrogen
Characeae	*Chara*, oogonium	1	1		Freshwater lakes, stagnant but oligotroph waters (submerged)
Characeae	*Chara*, oopore	1			Freshwater lakes, stagnant but oligotroph waters (submerged)
Cyperaceae	*Cyperus longus*	2			Hydrophyt, living in Phragmites-swamps, river shore, muddy river shore, sea coast, ditches
Potamogeton-aceae	*Potamogeton* cf. *nodosus*			3	Hydrophyt in slow flowing eutrophic waters (brooks)
Ranunculaceae	*Ranunculus sceleratus*			1	Nitrogen indicator on occasionally flooded muddy ground; very salt-tolerant
Rosaceae	*Rubus caesius*	2			Mainly by streams and rivers in sun or shade
Rosaceae	*Rubus* sp.			2	Probably close to river environments; prefers medium moist to moist soils, moderately enriched in nitrogen
Salicaceae	*Salix* sp., bud	1	1		Riverbanks
Solanaceae	*Solanum dulcamara*			1	Nitrogen indicator common in willow woods along riverbanks on moist to wet soils
Asteraceae	Asteraceae	3			-
Cyperaceae	*Carex* spp.	3			-
Cyperaceae	Cyperaceae			2	-
Euphorbiaceae	*Euphorbia* sp.	1			-
Lamiaceae	*Ajuga/Teucrium*	1			-
Lamiaceae	Lamiaceae	1			-
Polygonaceae	*Rumex* sp.			23	-
Rosaceae	*Rubus* sp., thorns	3			-
Solanaceae	*Solanum* sp.			1	-
	indet. plant remains			7	
	Sheep/goat dung pellet			1	

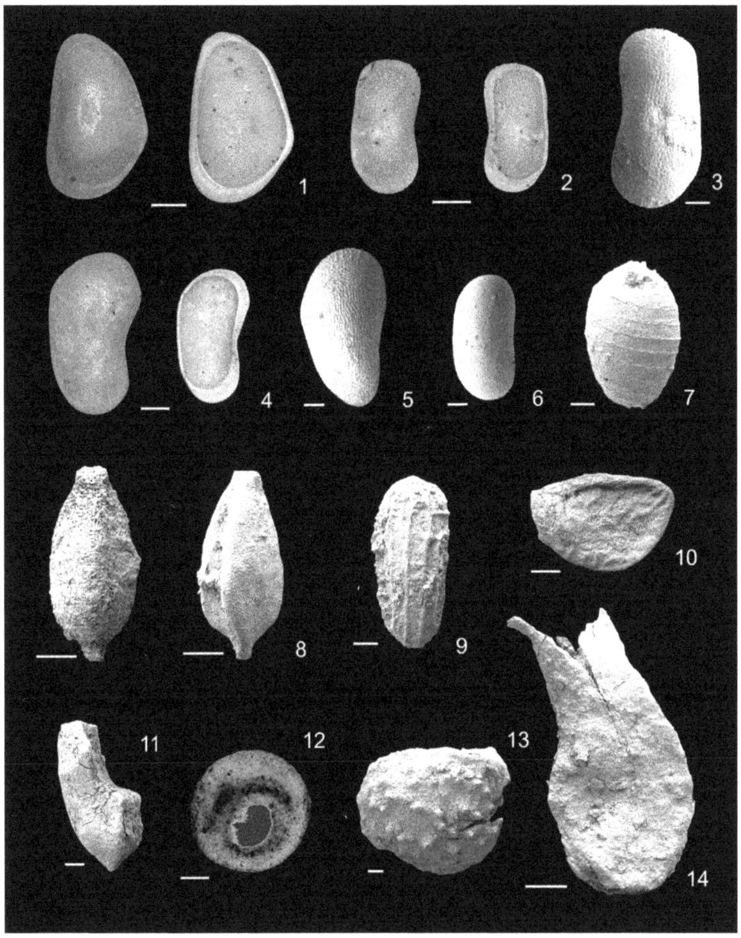

Fig. 3: Ostracoda and plant remains from HAM sequences (scale dimension in parentheses). Ostracoda: 1. *Prionocypris zenkeri* (250 µm); 2. *Ilyocypris* cf. *bradyi* (250 µm); 3. *Ilyocypris inermis* (100 µm); 4. *Candona* cf. *neglecta*-group (250 µm); 5. Cyprididae (cf. *Bakunella* sp.; 100 µm); 6. *Candona* sp. (100 µm). Botanical remains: 7. *Chara* sp. (100 µm); 8. *Cyperus longus* (200 µm); 9. *Verbena officinalis* (200 µm); 10. *Rubus* sp. (500 µm); 11. *Triticum dicoccum*, glume base fragment (200 µm); 12. *Galium* sp. (500 µm); 13. *Ranunculus arvensis* (200 µm); 14. *Salix* sp., bud (500 µm).

measurements for the Middle Bronze Age and thus would indicate a considerable stress signal for the end of the Chalcolithic.[16]

In any case, high frequency of barley, free-threshing wheat, and fruit trees indicate a change in agricultural and economic management with the transition from the Chalcolithic to the EBA.

In the organic layer of the HAM sequence, emmer wheat chaff is the most abundant find amongst the carbonised crop remains, followed by barley grain. Given the comparatively high frequency of emmer wheat and the underrepresentation of fruit trees in Chalcolithic sites of the region, the composition of the HAM samples may fit well into the Chalcolithic period.

3.3 Ecological classification of the local area around 3450 BC

Considering the probable weeds of the HAM assemblage, which are represented by most of the carbonised wild plant species, their water requirements allow growth on dry to medium-moist soils that are slightly enriched in nitrogen and intolerant to salinity. This reveals that their cultivation took place further away from the river banks.

[16] Riehl *et al.* 2007.

In general, two ecological units are represented in the uncarbonised assemblage:

1. Open vegetation, which includes disturbed habitats, ruderal, and weed species; and
2. Hydrophytic vegetation.

While the open vegetation is almost equally composed of ruderals and typical weed species, the representatives of the hydrophytic vegetation are diverse and indicate the presence of at least three different plant associations. These are, according to Zohary,[17] the Potametea, the Phragmitetea, and the Populetea associations.

The Potametea class is localised at the waterside belt of the Phragmitetea class and includes the submerged *Chara* algae and *Potamogeton* species. Their presence indicates slow flowing waters, rather than high stream velocity. *Chara* prefers oligotroph water conditions.

The Phragmitetea, which follows the Potametea class landwards, includes many high growing grasses. One such example is the medium high-growing *Cyperus longus,* which can be a dominant species in such units. In river bends that have more stagnant water and high nitrogen content, other species like *Ranunculus sceleratus* may have also grown.

The Populetea class follows the Phragmitetea and contains the tree genera *Populus* and *Salix*. The latter is present in the form of buds within the HAM assemblage. Other species that may have grown in this association are *Rubus* species, *Vitis* sp., *Solanum dulcamara*, and *Verbena officinalis.*

In all, the riverine vegetation was rich around 3450 BC due to a steady flow of the Jaghjagh, which also enabled the development of diversified underwater vegetation. In consideration of the transversal extension of the organic layer for at least 21.5 m, the appearance of this vegetation type may have been similar to that still present today along the central part of the Jordan River, north of the discharging Jabbok River. Unfortunately, nothing can be said about the extension of the ecotone.

The composition of the ostracod fauna particularly complements the specific water conditions shortly before the organic layer was covered.

The ostracod composition of the samples differs. While HAM B3 is dominated by *Ilyocypris* cf. *inermis* and representatives of the *Candona* cf. *neglecta* group, HAM III is mainly composed of *Prionocypris zenkeri* and *Ilyocypris* cf. *bradyi.* Only the species of *Candona neglecta* are present in both locations. These are ecologically characterised as tolerating a wide range of aquatic habitats, but prefer cold waters. *Ilyocypris* cf. *inermis* is able to tolerate slightly brackish conditions[18] and can populate the mud at the bottom of springs. This indicates the slight local differences within the river ecosystem, with the species of HAM B3 populating a more shallow area. Furthermore, *Prionocypris zenkeri* avoids deep as well as stagnant waters and prefers waters with rich aquatic vegetation.

Although there may be slight autecological differences in the different species, they all are symptomatic of cool waters, flowing with moderate to slow velocity, with a preference of springs and ponds by springs. This indicator quality derived from all the species best describes the characteristics of the Jaghjagh River around 3450 BC.

4. Conclusions

The analysis of plant remains and ostracods from an alluvial organic layer enabled the reconstruction of the riparian vegetation, which contains components of Potametea, Phragmitetea, and Populetea, and the hydromorphology of the middle Jaghjagh at around 3450 BC.

The palaeobiological investigation of alluvial sediments in modern semi-arid and arid environments provides valuable information on ancient river ecology, hydrography, and hydrology, which are essential in the reconstruction of landscape development and, thus, for an understanding of ancient population dynamics. The latter of which further elucidates the environmental basis of economic development during the time period under consideration.

Bibliography

ALTINSACLI, S. – GRIFFITHS, H.I.
2002 A review of the occurrence and distribution of the recent non-marine Ostracoda (Crustacea) of Turkey, Zoology in the Middle East 27, 61-76.

---

[17] Zohary 1973.

[18] Altinsacli – Griffiths 2002.

BERTSCH, K.
1932 Die Pflanzenreste der Pfahlbauten von Sipplingen und Langenrain im Bodensee, Badische Fundberichte 2.

DECKERS, K. – RIEHL, S.
2007 Fluvial environmental contexts for archaeological sites in the Upper Khabur basin (northeastern Syria), Quaternary Research 67, 337-348.

ELLENBERG, H. – WEBER, H.E. – DÜLL, R. – WIRTH, V. – WERNER, W. – PAULISSEN, D.
1991 Zeigerwerte von Pflanzen in Mitteleuropa, Scripta Geobotanica 18.

EVANS, J. – O'CONNOR, T.
2001 Environmental archaeology – Principles and methods.

FREELS, D.
1980 Limnische Ostrakoden aus Jungtertiär und Quartär der Türkei, Geologisches Jahrbuch, Reihe B 39.

GRIFFITHS, H.I. – HOLMES, J.A.
2000 Non-marine ostracods and Quaternary palaeoenvironments, Quaternary Research Association Technical Guide 8.

GRIFFITHS, H.I. – SCHWALB, A. – STEVENS, L.R.
2001 Environmental change in southwestern Iran: the Holocene ostracod fauna of Lake Mirabad, The Holocene 11(6), 757-764.

HALD, M. – CHARLES, M.
2008 Storage of crops during the fourth and third millennia B.C. at the settlement mound of Tell Brak, northeast Syria, Vegetation History and Archaeobotany 17(1), 35-41.

HARTMANN, G.
1964 Asiatische Ostracoden, Systematische und zoogeographische Untersuchungen, Internationale Revue der gesamten Hydrobiologie, Systematische Beihefte 3.
1989 Ostracoda, Klassen und Ordnungen des Tierreichs 5.

JENSEN, H.A.
1998 Bibliography on seed morphology.

KEMPF, E.K.
1973 Ostrakoden aus dem Jungpleistozän des Damaskus-Beckens, Zeitschrift für Geomorphologie 17 (3), 336-347.

LOWE, J.J. – WALKER, M.J.C.
1997 Reconstructing quaternary environments.

MARTENS, K. – ORTAL, R.
1999 Diversity and zoogeography of inland-water Ostracoda (Crustacea) in Israel (Levant), Israel Journal of Zoology 45, 159-173.

MCCORRISTON, J. – WEISBERG, S.
2002 Spatial and temporal variation in Mesopotamian agricultural practices in the Khabur Basin, Syrian Jazira, Journal of Archaeological Science 29 (5), 485-498.

MEISCH, C.
2000 Freshwater ostracoda of Western and Central Europe.

MOUTERDE, P.
1966 Nouvelle flore du Liban et de la Syrie.

PACHUR, H.J. – HOELZMANN, P.
1991 Paleoclimatic implications of late Quaternary lacustrine sediments in Western Nubia Sudan, Quaternary Research (Orlando) 36 (3), 257-276.

POST, G.E.
1932 Flora of Syria, Palestine and Sinai, 1 and 2.

RIEHL, S. – KÜMMEL, C.
2005 Archaeobotanical database of Eastern Mediterranean and Near Eastern sites. http://www.cuminum.de/archaeobotany (date referred to).

RIEHL, S. – FERRIO, J.P. – WILLCOX, G.
2007 Stable carbon isotope evidence of changing water stress on cereal crops in ancient agricultural societies in the Near East (5300-2000 BP), Poster presented at the SIMSUG 2007, Newcastle upon Tyne.

RIEHL, S. – PUSTOVOYTOV, K.E. – HOTCHKISS, S. – BRYSON, R.A.
2008 Local Holocene environmental indicators in Upper Mesopotamia: pedogenic carbonate record vs. archaeobotanical data and archaeoclimatological models, Quaternary International, 209 (1-2), 154-162.

VAN MORKHOVEN, F.P.C.M.
1963 Post-Palaeozoic Ostracoda, Their morphology, taxonomy, and economic use.

WIRTH, E.
1971 Syrien: Eine geographische Landeskunde.

ZOHARY, M.
1973 Geobotanical foundations of the Middle East.

9. Tell Brak's loess section in TC

Tony J. Wilkinson and Katleen Deckers

1. Introduction

Tell Brak is located in the southern Upper Khabur, an area that presently receives an annual rainfall of less than 300 millimetres (fig. 1). Today, eolean activity is common here but due to continued cultivation and ploughing, at most places eolean layers are insignificant and particle sizes are somewhat more poorly sorted than typical loess, containing larger percentages of fine fractions. However, in more sheltered areas, recent eolean accumulation has been observed with a more typical loess spectrum.[1]

Tell Brak was one of the largest urban centres of northern Mesopotamia and was originally excavated in 1937-1938, and again since 1976. The site is formed of major occupation horizons dating between the sixth millennium and end of the second millennium BC.[2] In excavation area TC, above cultural layers, natural deposits of considerable thickness were found that deserved further investigation.

The described section in Trench TC of Tell Brak was excavated in 2000 under the direction of Geoff Emberling and Helen McDonald (fig. 2). The section examined in September 2006 consisted of a vertical 3.8 metre-high upright soil section located at the south end of Trench TC, away from the so-called "Brak Oval."[3] Artefacts were rare to absent throughout the sequence, although occasional sherds occurred in the lowest level (D, as described here). In the field it was conjectured that the upper stratum may have been disturbed by bulldozing because the very top part of the sequence appears to have been lost. Due to the fact that these horizons apparently post-date the Brak oval complex (of an Early Dynastic date), as well as the Akkadian levels above them, stratigraphically they can be regarded as having a post-Akkadian date.[4]

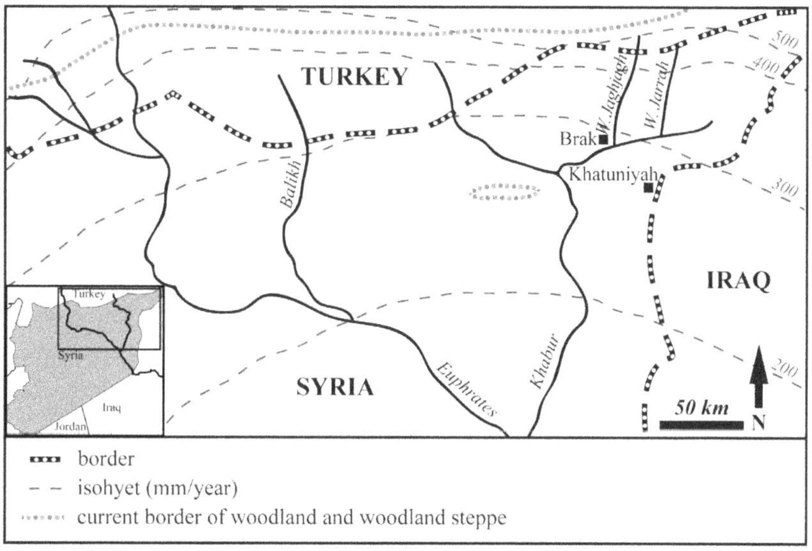

Fig. 1: Location of Tell Brak.

[1] Rösner 1995, 177-178.
[2] e.g. Oates – Oates 1994.
[3] Emberling – McDonald 2001, 31-45.
[4] Emberling – McDonald 2001, 35.

2. Field descriptions (fig. 3):

A) 0-110 centimetres. Fine sandy silt loam below a cleared surface. Weak boundary at base of horizon. Eolian deposit.

B) 110-160 centimetres. Soft pale greyish brown silt loam; heavily bioturbated with common-to-abundant small biopores and occasional larger animal burrows. Merging boundary below. Appears to represent a weak palaeosol horizon and a period of stability.

C) 160-300 centimetres. Thick accumulation of pale brown silt loam with occasional-to-common vertical pores from fine roots. Between one and five fine laminae of silt or silty clay appear to represent deposits from low-energy wash or localised bodies of water. These may represent phase(s) of seasonal washing of water into a shallow depression.

D) 300-380 centimetres. Pale brown sandy loam to sandy silt loam. Contains more sand and grit than horizon C. Also includes occasional artefacts, which suggests that deposit D has been resorted from nearby archaeological deposits.

E) Not in section but exposed to the east of the trench. Archaeological deposits dipping into basin.

Overall, the sediments consist throughout of fine sandy silt, loess-type deposits. The slightly darker bioturbated horizon B appears to represent a weak phase of stability during which there was sufficient time for soil biological activity to occur. The thin laminae of silt in C, together with the presence of vertical fine root passages, suggests that the accumulation of eolian silt was interrupted by occasional episodes of either soil wash or puddle formation.

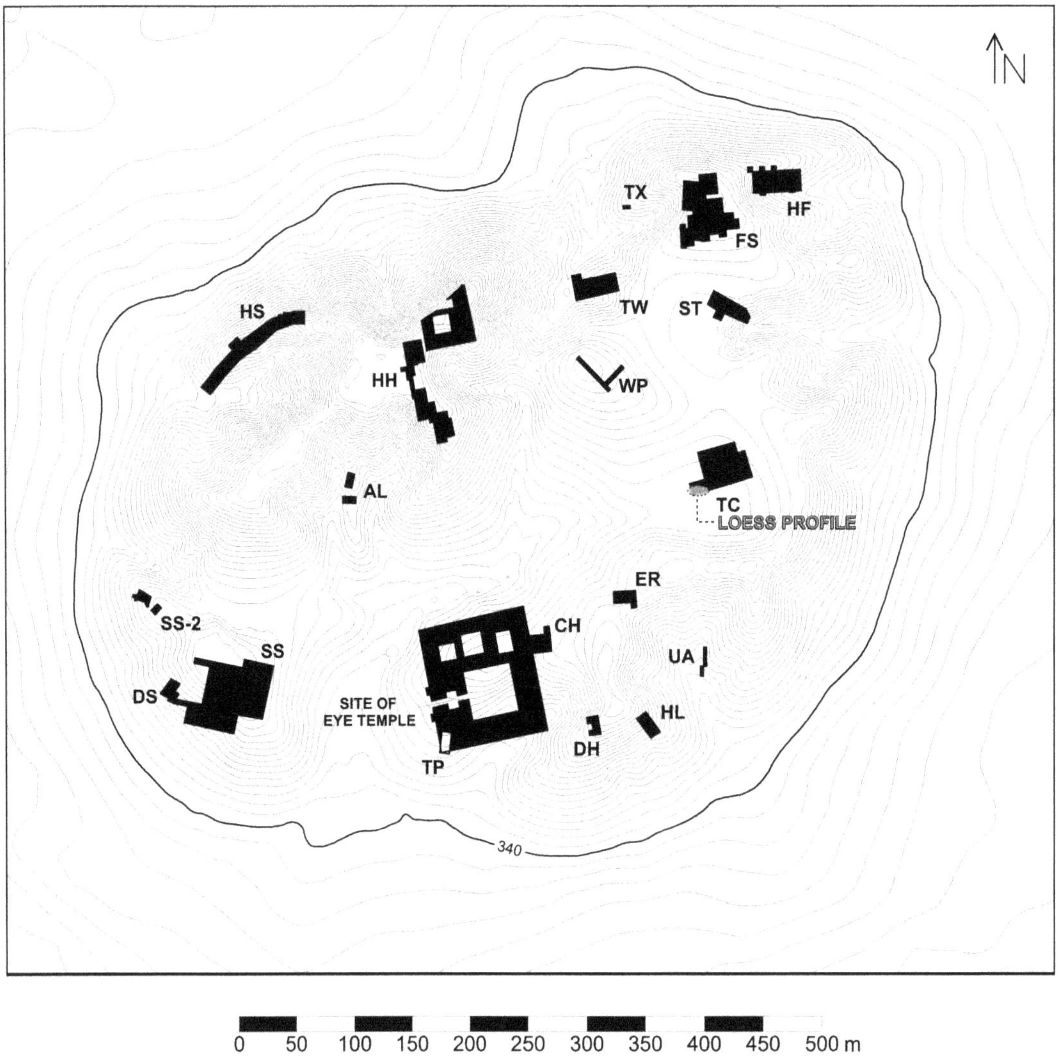

Fig. 2: Location of area TC, where the profile was investigated (map by Tim Skuldboel).

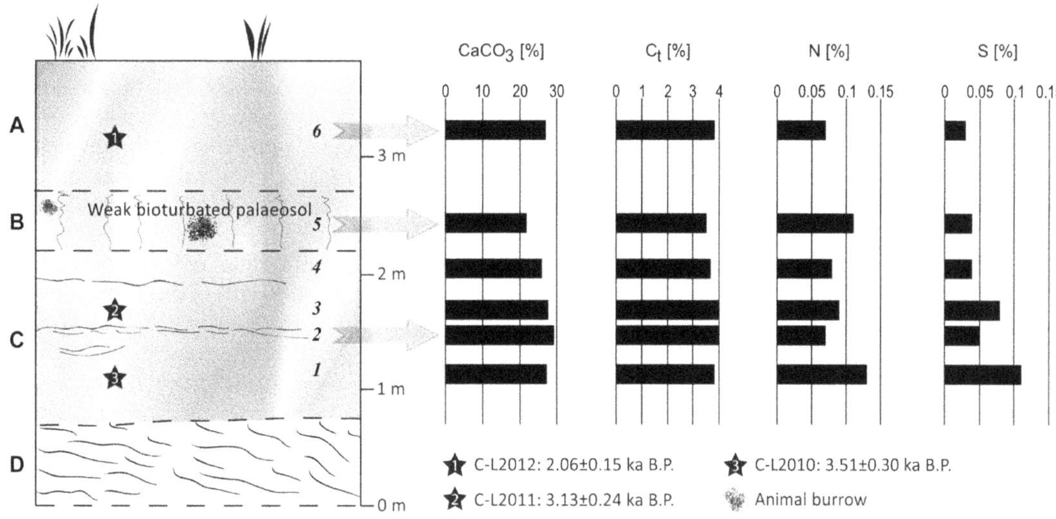

Fig. 3: Illustration of loess profile with sedimentological laboratory results.

The only sediments with direct evidence for accumulations in the presence of human occupation are in E, located immediately to the east of the section. Level D appears to have accumulated as a result of wind deposition, albeit in proximity to archaeological horizons.

3. Sedimentological analysis

Six sediment samples were collected to gain increased insight into depositional and soil formation processes. Sedimentological analysis on the bulk samples was undertaken by the laboratory of the Department of Geography, University of Tübingen. Calcium carbonate content, pH, and particle size analysis were measured according to the procedures described in Beck *et al.*[5] Carbon, nitrogen, and sulfur percentages were measured using the Vario ElIII elemental analyser. Additionally, three sediment samples were retrieved for optically stimulated luminescence dating from the section at TC (see fig. 2) and were dated by Dr. A. Hilgers.[6] The lowermost date was 3.51 ± 0.3 kyr (C-L2010), the one above 3.13 ± 0.24 kyr (C-L2011), and the uppermost one 2.06 ± 0.15 kyr (C-L2021), hence providing an internally consistent sequence (fig. 3).

Particle size analysis results confirm that the largest proportion of the samples consists of medium silt. However, a fair proportion of clay also is present within the samples; with especially higher clay percentages in the upper half of the profile. Additionally, a smaller sand-sized fraction was identified. These might, however, represent pseudo-sand, and have developed either by soil formation processes or as a result of the deflation of silt-clay aggregates. Significant amounts of micro-charcoal are present in all samples. This partially explains the rather high total carbon percentages, at between 3.5 and four throughout the profile (fig. 3). However, calcium carbonate percentages are also high – above 25% in all samples; and this also contributed to the relatively high carbon percentages. Carbon and sulfur percentages are somewhat higher in the lower part of the section and no particular carbon or sulfur peak was observed for sample 5, retrieved from the palaeosol. Calcium carbonate percentages were also somewhat lower at this level, which may account for the total lower carbon percentages (fig. 3).

4. Discussion

In terms of chronology, field evidence suggested that the entire sequence had accumulated as a result of eolian deposition; the latter occurring because of the sheltered conditions that prevailed in this broad, open basin. The weak palaeosol (B) appears to represent a phase of stability of indeterminate duration that is marked with some biological activity; perhaps weak accumulation of organic soil matter and increased activity of small burrowing animals. The OSL dates suggest that, overall, the accumulation of sediments was fairly steady at around 1.2 – 1.6 millimetres per annum, which is equivalent to 1.2 – 1.6 metres per millennium. This roughly accords with the date of the archaeological deposits overlaid by this loess (ca. 4300-4400 BP) and the overall depth of the section, 3.8 metres. The slight decrease in the sedimentation rate in

[5] 1995.

[6] see chapter 11 this volume.

the upper two metres may be partly accounted for by the episode of stability represented by the palaeosol, which must have been associated with decreased silt accumulation rates and increased pedogenesis.

Overall, this sedimentary sequence suggests that the accumulation of eolian loess was both relatively steady and continuous over the four millennia since the building complex at TC was abandoned. Although one should not draw too many conclusions from only three OSL dates, there is no evidence for any episode of accelerated dust-fall such as has been postulated for the final centuries of the third millennium BC.[7] Additionally, the only real evidence for changing rates of sedimentation derives from the slight decrease between 3130 and 2060 BP, and the palaeosol in B. This may correspond roughly to the hiatus of occupation that occurred after the Middle Assyrian occupation in HH (on the summit of the tell) and the resumption of occupation in the vicinity of the "castellum" during the Partho-Roman and Sasanian Periods (i.e. between ca. 1200 BC and the 1st century AD).[8] Because occupation at Tell Brak would have been associated with increased agricultural activity in the surrounding fields, it is possible that the dust-fall relates to the increased levels of eolian dust kicked up by local agricultural activity as well as settlement on and around the tell.

Interestingly, Rösner[9] found similar evidence of increased eolean sedimentation between 2.17 ±75 and 2.04 ± 80 kyr BP, on the plain south of the Wadi Jaghjagh towards Khatuniyah, when comparing 2.04 ± 80 kyr BP to today. There, sedimentation rates of 0.4-0.87 centimetres per annum took place in the first period and then decreased to 0.06 centimetres per annum afterwards. Rösner[10] also concludes that the increased eolean activity at this time is anthropogenically influenced. Overall drier climatic conditions between roughly 3500 and 1000 BP, as visible in the oxygen isotopic record of Lake Van, may however also have played a role in the increased eolean activity in this region.[11]

Bibliography

EMBERLING, G. – MCDONALD, H.
2001 Excavations at Tell Brak 2000: Preliminary report, Iraq 63, 21-54.

OATES, D. – OATES, J.
1990 Aspects of Hellenistic and Roman settlement in the Khabur Basin, in: P. Matthiae – M. van Loon – H. Weiss (eds.), Resurrecting the past: a joint tribute to Adnan Bounni, 227-248.
1994 Tell Brak: a stratigraphic summary, 1976-1993, Iraq 56, 167-76.

RÖSNER, U.
1995 Zur quartären Landschaftsentwicklung in den Trockengebieten Syriens.

WEISS, H. – COURTY, M.-A. – WETTERSTROM, W. – GUICHARD, F. – SENIOR, L. – MEADOW, R.A. – CURNOW, A.
1993 The genesis and collapse of third millennium north Mesopotamian civilization, Science 261, 995-1004.

WICK, L. – LEMCKE, G. – STURM, M.
2003 Evidence of Late Glacial and Holocene climatic change and human impact in eastern Anatolia: High-resolution pollen, charcoal, isotopic and geochemical records from the laminated sediments of Lake Van, Turkey, The Holocene 13.5, 665-675.

---

7 Weiss *et al.* 1993.

8 Oates – Oates 1990; 1994.

9 1995, 115.

10 1995.

11 Wick *et al.* 2003.

10. Luminescence dating of fluvial deposits near archaeological sites in the Upper Khabur basin (NE Syria): an introduction with illustrative applications

D.A.G. Vandenberghe[a, b]

1. Introduction

For a long time now, water availability has governed human activity in the Upper Khabur basin of NE Syria. Indeed, many archaeological settlements (often tells, or settlement mounds) are located along the tributaries of the Upper Khabur.[1] Man, in its turn, has largely affected the appearance of the landscape. Damming and irrigation caused some water courses to run dry in summer, and deforestation, grazing and cultivation resulted in a landscape that is, at present, devoid of trees. Because of anthropogenic forcing, the present-day environmental conditions in the Upper Khabur basin do not accurately reflect those prevailing at the time of archaeological occupation. To improve our understanding of the development and functioning of these settlements, it is necessary to unravel their former environmental settings. This requires a chronological framework for the regional evolution of the fluvial landscape and climate.

In this book, optically stimulated luminescence (OSL) dating is one of the methods that has been applied to establish an absolute chronology for the evolution of the riverine environment around several well-studied archaeological sites in the Upper Khabur basin. This contribution summarises the work that has been accomplished by the author to this respect. Rather than providing a pure technical report, an attempt is made at making the research results more accessible to the non-specialist. This paper, therefore, first outlines the general principles of luminescence dating, as well as its most important possibilities and limitations (Section 2). Throughout, the focus is on quartz, as this is the dosimeter that has been used. The introduction to luminescence dating has largely been adopted from Vandenberghe[2] which, in its turn, builds on some of the key texts and reviews listed in Appendix A. The paper then reports on the application of OSL dating to fluvial deposits in NE Syria (Section 3). The OSL characteristics of fluvial quartz grains in the study area are documented, and a summary is given of the ages based on these signals. For some sites, sufficient age information is available to allow a discussion on the reliability and/or potential of optical dating in this environment.

2. Luminescence dating

2.1 General principles

Luminescence dating belongs to the category of *radiation dosimetry dating methods* (also called *radiation damage methods*), which are based on the time-dependent accumulation of radiation damage in minerals. Other members belonging to this family of dating methods are fission-track dating and electron spin resonance dating. The radiation damage is the result of the exposure to a low-level of ionising radiation that is omnipresent in nature. The longer a mineral is exposed to radiation, the greater the radiation damage in its crystal lattice. The amount of radiation damage is consequently a measure for the total dose of radiation energy absorbed by the mineral throughout time.

In luminescence dating, the intensity of the radiation damage is detected as a small amount of light, which is called *luminescence*. The radiation damage, and hence the latent luminescence signal, is removed, or set to zero, by exposure to heat or light. For ancient pottery, for instance, the 'zeroing' took place during manufacturing, when it was baked in an oven. In the context of sediment dating, the zeroing event was the exposure to daylight during erosion, transport and deposition of the mineral grains. This zeroing through exposure to sunlight is also called *bleaching*. Once the mineral grains become incorporated in sediment and buried under other grains, the clock starts ticking anew.

---

[a] Laboratory of Mineralogy and Petrology (Luminescence Research Group), Department of Geology and Soil Science, Ghent University, Krijgslaan 281 (S8), B-9000 Gent, Belgium - email: dimitri.vandenberghe@UGent.be

[b] Postdoctoral Fellow of the Research Foundation – Flanders (FWO – Vlaanderen)

[1] Deckers – Riehl 2007.

[2] 2004.

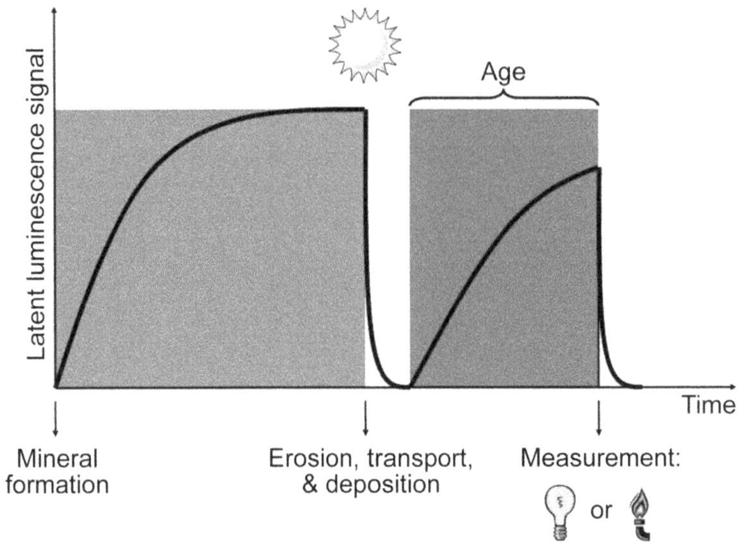

Fig. 1: Schematic representation of the event that is being dated in the luminescence dating of sediments. Minerals are continuously exposed to a low-level of natural radioactivity, through which they can acquire a latent luminescence signal. During erosion, transport and deposition, the minerals are exposed to sunlight and all the previously accumulated luminescence is removed ("bleaching"). Once shielded from the sunlight, the signal starts to build up again, until the moment of measurement in the laboratory. The age that is being determined is consequently the time that has elapsed between these two zeroing events. The luminescence is measured in the laboratory by exposing the mineral grains to light or heat, and is then termed *optically stimulated luminescence* (OSL) or *thermally stimulated luminescence* (thermoluminescence or TL), respectively.

The latent luminescence signal can also be released in the laboratory using the same zeroing agents, heat or light. At this point, however, the luminescence signal that is being emitted will be recorded. If the signal is set free by applying heat, the resulting luminescence is called *thermoluminescence* (TL). If the measured signal results from exposure of the mineral grains to a beam of light, the emitted luminescence is termed *optically stimulated luminescence* (OSL).

The luminescence signal measured in the laboratory is thus related to the dose the mineral has received since the last zeroing event. For sediments, the last zeroing event was their exposure to sunlight at deposition. If the rate is also determined at which the dose was absorbed (or, in other words, the radiation damage was created), then it is possible to calculate an age. This age refers to the time that has elapsed between the moment that a sediment was deposited and the moment of sampling for analysis. This is illustrated in Figure 1.

2.2 Mechanism

Although the reality is much more complicated and in fact poorly understood, the main processes causing luminescence can be described in terms of the energy band model for non-conducting ionic crystalline materials (fig. 2).

In this model, electrons are associated with discrete ranges of energy, which are called *bands*. The lowest energy band is the valence band; the highest energy band is the conduction band. The gap between the two is the so-called 'forbidden zone'. In an ideal crystal lattice, no electron occupies a position in this zone. In any natural crystal, however, there are defects that disturb the perfectly ordered crystalline structure. Many types of imperfections are possible such as impurities (either as substitutionals or in interstitial positions) or missing atoms (vacancies). These defects give rise to localised energy levels within the forbidden zone. These localised energy levels are the key to the luminescence phenomenon, as they carry the memory of exposure to nuclear radiation. In other words, luminescence requires the existence of lattice defects.

In every rock or sediment there is a low level of nuclear radiation. This radiation has an ionizing effect and, upon interaction with the crystal, can raise an electron from the valence band into the conduction band

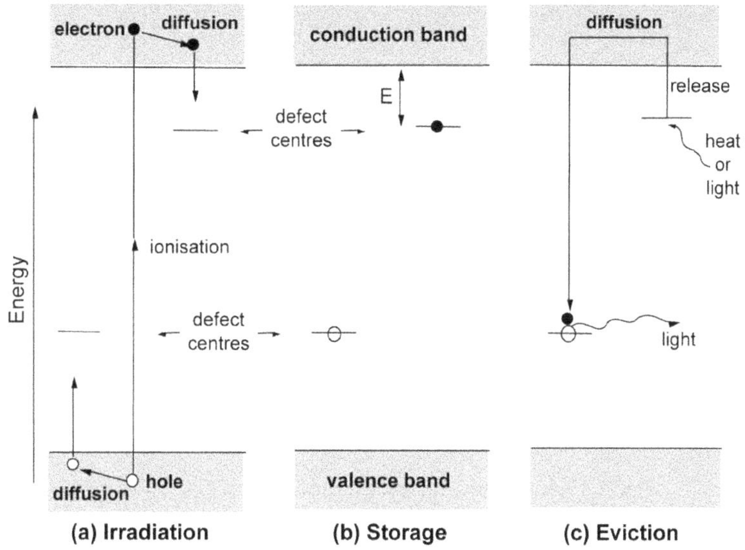

Fig. 2: Energy-level representation of the luminescence process.
○: hole; •: electron.

(fig. 2a). For every such electron that is generated, an electron vacancy, a *hole*, is left behind and both the electron and hole are free to move through the crystal. In this way, energy of the nuclear radiation is taken up. The energy can be released again (usually as heat) by recombination. Although most charges indeed do recombine directly, it also happens that the electron and the hole are trapped at the defect centres. In this case, the nuclear energy is stored temporarily in the crystal lattice and the system is said to be in a metastable situation (fig. 2b). Energy is required to remove the electrons out of the traps and to return the system to a stable situation. The amount of energy that is necessary, is determined by the depth E of the trap below the conduction band. This trap depth consequently determines how long an electron will stay in the trap. To empty deeper traps, more energy will be required and those traps are more stable over time. For dating, we are only concerned in traps that are deep enough (i.e. ~ 1.6 eV or more) for the lifetime to be at least several million years.

By exposing the crystal to heat or light, the trapped electrons are evicted out of their metastable positions (fig. 2c). Once evicted, they can recombine with holes in the so-called *recombination centres*. These are defect sites attractive to electrons. This recombination can result in the emission of light (radiative recombination). The defect sites where the radiative recombination occurs are called *luminescence centres*, and the resulting light is TL or OSL, depending on whether heat or light, respectively, was used to release the electrons from the traps. The colour (wavelength) of the emitted luminescence depends on the type of luminescence centre where the recombination takes place. The amount of light emitted is proportional to the amount of electrons stored in defects, and hence to the amount of energy absorbed from nuclear radiation. Since the energy is absorbed at a certain rate, the intensity is related to the time of accumulation; the longer the material is exposed, the more signal is acquired. At some point, however, the luminescence signal reaches a saturation level where it no longer increases with radiation dose.

2.3 Stimulation of the signal

In TL, the trapped electrons are released by heating the sample at a fixed rate from room temperature to between 450°C and 700°C. As shown in Figure 3, the signal is plotted as a function of the measurement temperature; the resulting plot is referred to as a *glow curve*. The TL signal observed at higher measurement temperatures originates from traps that are deeper below the conduction band (and, therefore, are more stable). In OSL, the electron escapes from its trap as the result of the absorption of a photon of light with a sufficient energy. As soon as the stimulating light is switched on, luminescence is emitted by the mineral grains. The signal is recorded as a function of time and the resulting plot is called a *shine-down* or *OSL decay curve* (fig. 4). As the measurement continues, the electrons in the traps are emptied and the signal decreases.

Optical stimulation has now become the most commonly used method for sediment dating. The light used to stimulate the minerals is restricted to a narrow range of (typically longer) wavelengths so that it can be well separated from the emitted luminescent light (of typically shorter wavelength). The type of mineral that is under investigation plays an important role in the selection of the most appropriate light source for stimulation. Quartz is usually stimulated by visible light with a wavelength in the blue to green region of the spectrum. For feldspars, long wavelengths in the infrared region (800-900 nm) can also be used.

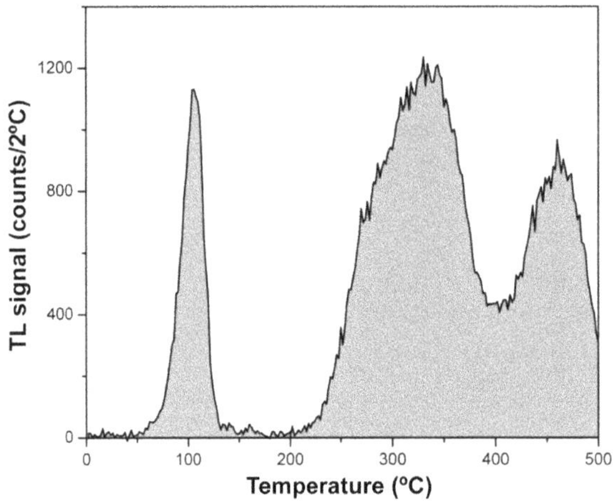

Fig. 3: Example of a quartz TL glow curve. The signal was induced by irradiation in the laboratory, which explains why TL appears at temperatures <200°C.

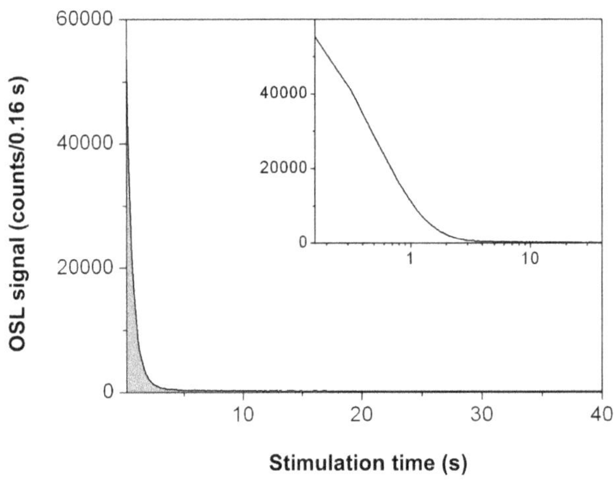

Fig. 4: Example of a quartz OSL decay curve. The inset shows the same data on a semi-log scale.

Depending on the wavelength used for stimulation, the resulting luminescence is termed *infrared stimulated luminescence* (IRSL), *blue-plus-green stimulated luminescence* (BGSL), *green light stimulated luminescence* (GLSL or GSL), or *blue light stimulated luminescence* (BLSL or BSL).

2.4 Minerals used for dating

Both feldspars and quartz have been found to emit OSL signals and they are the most widely used minerals in the optical dating of sediments. The choice of the mineral that is being used usually depends on the availability within a sample, and on the age of the sediments. Quartz saturates at lower doses than feldspars, and so the use of feldspar might prove to be advantageous for dating older deposits. A common problem encountered with feldspars, on the other hand, is anomalous fading of the natural signal, which leads to age underestimation.[3] Methods to measure and correct for anomalous fading have been suggested,[4] but their accuracy remains to be established.

Whereas for sand-sized grains (of ~0.1 mm in diameter) the different mineral fractions can be easily separated from each other, this is not the case for silt-sized grains (of ~0.01 mm in diameter). Traditionally, the measurements are then carried out on the polymineral fine fraction, which is a mixture of all kinds of

[3] Wintle 1973.

[4] Huntley – Lamothe 2001; Lamothe *et al.* 2003; Kars *et al.* 2008.

minerals with a grain size within the range 4-11 μm. It is possible to extract fine-grained quartz from the polymineral fraction through an appropriate treatment with acid (such as H_2SiF_6), and this approach is being increasingly used in the dating of sediments that are predominantly composed of silt-sized grains.[5]

2.5 The age equation

A luminescence age is obtained by solving the equation:

$$\text{Age (a)} = \frac{\text{Burial dose (Gy)}}{\text{Annual dose (Gy a}^{-1}\text{)}}$$

The burial dose is the energy absorbed from nuclear radiation by the sedimentary mineral grains since they were last exposed to sunlight. This quantity is determined using luminescence signals as an equivalent dose (ED or D_e), which is the artificial laboratory dose necessary to induce a luminescence signal in the minerals that is of exactly the same magnitude as their natural signal (see Section 2.7). The D_e is usually determined using β-radiation. The unit of absorbed dose is the Gray (symbol: Gy) and is defined as 1 Joule per kilogram.

The annual dose or dose rate, on the other hand, refers to the radiation energy that is annually absorbed per unit of mass and consequently it has the unit of Gy a^{-1}. The radiation comes from the naturally occurring nuclides of the ^{232}Th, ^{238}U and ^{235}U decay series, from the decay of ^{40}K and ^{87}Rb, and from outer space (cosmic radiation). The annual dose is usually derived from the measurement of elemental or radionuclide concentrations (see Section 2.6).

2.6 The dose rate

2.6.1 Luminescence efficiency

The decay of the naturally occurring radionuclides is attended by the emission of alpha particles, beta particles and/or gamma rays. ^{40}K, for example, produces β and γ radiation, while the ^{232}Th, ^{238}U and ^{235}U decay series produce α, β and γ radiation. β and γ radiation have the same efficiency in inducing luminescence; meaning that the same luminescence signal results when an identical beta or gamma dose is administered. The luminescence efficiency of α radiation, however, is approximately 10 times lower. As the D_e is usually obtained using a β source, the alpha dose rate must be corrected for this lower efficiency; this efficiency correction factor is known as the a-value[6] or the b-value.[7]

2.6.2 Penetration range

The penetration range of the three types of radiation is very different: α-radiation penetrates about 20 μm in minerals such as quartz, β-radiation about 2 mm and γ-radiation about 30-50 cm. The short range of alpha and beta particles limits their contribution to the immediate location of the sample. Gamma rays, however, have a much longer range and consequently a much larger volume of sediment provides the gamma dose to a sample. This explains why it is preferable to collect samples from surroundings that are uniform in radioactivity within a radius of 30-50cm. Indeed, an "inhomogeneity", such as a piece of rock or an intercalated sediment layer, may contribute to the dose rate experienced by the sample if it lies within this range; if it differs significantly in radioactivity from the sediment and/or it attenuates radiation in a different manner, this may lead to erroneous estimates of the dose rate.

In luminescence dating, the difference in penetration range is exploited to facilitate the dose-rate calculations, namely by selecting only a certain grain size fraction for the luminescence measurements. The fine-grain technique uses sediment grains with a size of 4-11 μm. These grains are small enough to ensure complete penetration by all three types of radiation. In the quartz coarse-grain technique, the chosen grain size is typically around 100 μm. The outer layer of the grains can be etched away, so that the alpha contribution from external radioactivity is effectively removed. This leaves only contributions from gamma rays and a partly attenuated contribution from beta particles. It is usually a good approximation to assume that quartz is free of radio-elements and, hence, that the internal dose rate is negligible; in the case of K-feldspar, however, the internal beta contribution from ^{40}K and ^{87}Rb becomes increasingly important with grain size.

---

5 see e.g. Sugisaki *et al.* 2010; Timar *et al.* 2010.

6 Aitken 1985.

7 Bowman – Huntley 1984.

2.6.3 Methods of determination

Various methods can be used for determining the dose rate. A straightforward method is to determine the K, U and Th content of the bulk sample; among methods used are AAS (for K), ICPMS (for K, U and Th) or INAA (for K, U and Th). Many luminescence laboratories use techniques that directly measure the natural radiation emitted by the sediment sample, such as thick source alpha counting, beta counting and gamma-ray spectrometry. Of these, gamma-ray spectrometry is probably the method that is used most often nowadays. This technique measures the activities of one (or more) individual radionuclides in the U and Th decay chains, as well as ^{40}K. Gamma-ray spectrometry can be done in the field in auger holes in the sediment profile, or in the laboratory on bulk samples. On-site measurements are usually made using systems with a high detection efficiency but a limited energy resolution (such as a NaI(Tl)-detector); while this allows performing several measurements per day, the number of radionuclides that can be identified and quantified is too limited to give information about radioactive disequilibrium. Laboratory measurements are usually made with Ge-detectors. These systems have sufficient resolution for measuring multiple radionuclides in both the ^{238}U and ^{232}Th decay series so that information on the degree of radioactive equilibrium can be obtained. The measured concentrations, radionuclide activities or count rates are then converted to dose rates using factors derived from tabulated nuclear energy releases.[8] The contribution from cosmic rays is usually calculated using expressions that are available in the literature.[9]

Alternatively, the annual dose can also be directly determined (i.e. directly in terms of Gy/a) with the use of highly sensitive luminescent dosimeters. The principle of such an evaluation is in essence the same as for an age determination. A luminescent phosphor is first reset and subsequently buried in the sediment. It is left there for some time during which it is exposed to the natural radioactivity and hence absorbs a certain dose. This absorbed dose can then be determined using luminescence signals as outlined in Section 2.7. As the exact amount of time that the dosimeter was buried is known, the annual dose is then obtained by dividing the "equivalent dose" by this "age". The phosphor must be sufficiently sensitive to allow the measurements to be carried out on laboratory timescales. An example of a suitable phosphor is -Al2O3:C.

The various methods for dose-rate determination have their own advantages and disadvantages, so that, in some cases, it may be advantageous to use a combination of several techniques.

2.6.4 The effects of moisture

Water in the sediment pores absorbs part of the radiation that would otherwise reach the mineral grains; consequently, the dose-rate in sediments containing moisture is less than the dose rate measured in the same sediment when it is dry. If the dose rate is determined in the laboratory, the measurements are typically carried out on dry samples. To avoid age underestimation, the dry dose rates have to be corrected for the time-averaged moisture content. In case quartz grains are dated, an increase in water content of 1% generally corresponds to an increase in the optical age of about 1%. Fortunately, the effect is limited by the maximum amount of water a sample can take up (saturation). Uncertainty about the time-average moisture content, nonetheless, is a fundamental limitation in the precision that can be achieved.

Organic matter also attenuates radiation, and it does so differently from both water and the mineral grains; this needs to be considered, especially when organic-rich sediments are to be dated.[10]

2.6.5 Post-depositional processes

In the calculation of a luminescence age, it is normally assumed that the dose rate in the sediment has remained constant throughout time. Besides varying moisture content, there are several processes that may change the composition and porosity of a sediment after its deposition. Important processes in this respect are cementation and soil formation. As long as these processes do not significantly change the dose rate, the sedimentary constituent can be reliably dated; if such changes do occur, dating can become problematic. Another complication occurs if the members of ^{238}U series are not in secular equilibrium. Radon-222 (noble gas; half-life = 3.8 days) and Radium-226 (water soluble; half-life = 1600a) are the crucial isotopes in this respect. If these isotopes are able to escape from the sediment during its burial history, the radioactivity level measured today may not be representative of the level present, on average, during the burial period of a sediment sample. Laboratory high-resolution gamma-ray spectrometry allows for the measuring of several individual radionuclides in the uranium and thorium decay series and thus enables evaluating whether or not there is secular equilibrium; as it is not usually known when the disequilibrium set in, it is also not possible to correct for the effect.

---

8 see e.g. Adamiec – Aitken 1998.

9 Prescott – Hutton 1994.

10 Lian *et al.* 1995.

2.7 The equivalent dose

While there have been only minor advances in improving the accuracy and precision of the determination of the annual dose over the past twenty years or so, the determination of the equivalent dose (D_e) has been truly revolutionised. Until the beginning of the century, most D_e's were estimated using many subsamples (aliquots); typically a few tens of aliquots were required, each composed of a few milligrams of grains. Nowadays, all the measurements can be made on a single aliquot of quartz or feldspar grains; the size of the aliquot can even be reduced to that of a single grain. These measurement protocols are called *single-aliquot protocols.* Single-aliquot protocols have several advantages: (1) they allow for the variation in luminescence characteristics exhibited by different aliquots, leading to an increased precision by which a D_e can be determined, (2) the uncertainty on an average D_e can be based on the spread observed for several independent measurements rather than on the internal uncertainty associated with a modelled bulk dose-response characteristic, and (3) the distribution of doses within a sample can be examined explicitly (which allows, for instance, testing whether the luminescence clock in all grains has been completely reset). The protocol that brought about the breakthrough is the single-aliquot regenerative-dose (SAR) protocol. It was originally developed for use with OSL signals from quartz,[11] but it has now been proven to be useful with both OSL and TL signals from a wide variety of synthetic and natural dosimeters, ranging from polymineral fine grains and K-feldspar[12] to salt[13] and even washing powder.[14]

A generalised SAR measurement sequence is outlined in Table 1. The procedure starts with the measurement of the natural OSL signal ($L_{i,\ i=0}$), after the sample received a thermal pre-treatment. This measurement removes most, if not all, of the luminescence that would also be removed by exposure to natural sunlight. Luminescence signals are then repeatedly induced, or regenerated, by administering known laboratory doses; each time, the response to the regenerative dose is measured ($L_{i,\ i=1,2,3,...}$), which once again removes the signal. The size of the regenerative doses varies throughout the experiment, and they are chosen to yield signals that bracket the natural OSL signal. All these measurements are performed on a single aliquot. This repeated use may change the sensitivity of the aliquot, in which the sensitivity is defined as the luminescence output per unit of dose. In the SAR protocol, these changes are explicitly monitored by measuring the response to a test dose (T_i) that is given after each measurement of L_i (i.e. after the measurement of the natural, and each regenerated OSL signal). The test dose is kept constant throughout the entire experiment. If there are no changes in sensitivity, the test dose will always induce a signal of the same magnitude. Most, if not all, samples exhibit sensitivity change, which is then reflected by a change in the intensity of the OSL response to the test dose. It is possible to correct for this change by dividing the OSL signals (L_i) by the corresponding test-dose responses (T_i). A plot of this ratio (L_i/T_i) versus regenerative dose yields a curve that characterises the growth of the signal with dose. This curve is known as a *sensitivity-corrected growth curve* or *dose-response curve.* From this curve, the unknown dose (i.e. the equivalent dose) can be deduced by interpolation of the sensitivity corrected natural signal (fig. 5).

A number of tests are available to see whether a given SAR protocol is applicable to a particular sample. A first test checks whether the sensitivity changes that occur throughout successive measurement cycles are accurately corrected for. This is accomplished by a repeated measurement of the response to a dose equal to one that was administered previously. If the sensitivity correction is working properly, the two sensitivity-corrected signals should give the same value and their ratio (the so-called *recycling ratio*) should be unity. Such a repeat measurement is routinely incorporated in the measurement sequence (see fig. 5, open triangle).

A second test checks whether the dose-response curve indeed starts at the origin. This is necessary because the thermal pre-treatment (preheat) can give rise to an OSL signal even if no dose has been administered. The test consists of measuring the response to a zero dose. This should give a zero signal, but in practise a small signal is observed owing to recuperation. If this signal is small (a few percents of the corrected natural signal), it can be safely ignored; if it is large, it may be possible to correct for this contribution, or the result may be rejected. To minimise recuperation, a high-temperature optical cleanout is performed at the end of each SAR cycle (Step 7 in Table 1).

A third test is a dose recovery test. In such a test, the SAR protocol is used to measure a known laboratory dose that was administered to an aliquot after optical bleaching at room temperature and prior to any thermal treatment. The applied laboratory dose mimics the natural dose, and any protocol that is used for equivalent dose determination should at least be able to determine (or "recover") this known dose. In spite

---

[11] Murray – Wintle 2000; 2003; Wintle – Murray 2006.

[12] Buylaert *et al.* 2007; 2008.

[13] Bailey *et al.* 2000.

[14] Thomsen *et al.* 2002.

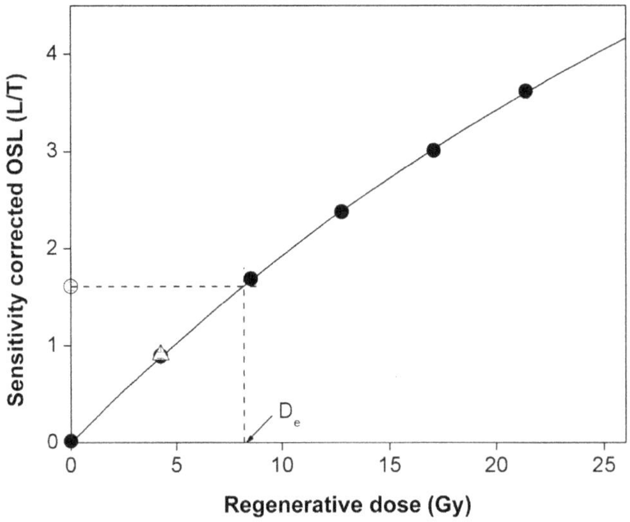

Fig. 5: SAR growth curve. The solid line is the fit of the data to a single saturating exponential function. The open triangle represents a recycling point. The equivalent dose (D_e) is obtained by interpolation of the sensitivity corrected natural OSL signal (open circle) on the corrected growth curve.

Table 1: Generalised SAR measurement sequence (Murray Wintle 2000). The optical stimulation at elevated temperature (step 7) is added to minimise recuperation (Murray Wintle 2003). For a natural sample, i = 0 and D_i is the burial dose.

Step	Treatment	Observe
1	Give dose D_i	
2	Preheat (10s at 160-300°C)	
3	Stimulate: 40s at 125°C	L_i
4	Give test dose D_t	
5	Preheat: ramp to 160°C*	
6	Stimulate: 40s at 125°C	T_i
7	Stimulate: 40s at 280°C	
8	Go back to Step 1	

* The sample is heated to a given temperature (160°C in this case) at a certain rate, and is cooled down as soon as this temperature is reached; this treatment is referred to as a *cutheat*.

of this, it is important to realise that although a good dose recovery is a minimum requirement, it does not guarantee that a naturally acquired dose is determined accurately as well. The test does allow evaluating whether some important boundary conditions are met, and the spread observed in a dose recovery experiment can be considered as a measure for the maximum precision that would be achieved in ideal conditions.

The measurement parameters (e.g. preheat, cutheat and stimulation time and temperature) in a SAR protocol may be found to vary from study to study, and the protocol can be used with different luminescence signals (i.e. OSL, IRSL, TL) from a variety of dosimeters (e.g. quartz, K-feldspar). What characterises a SAR-protocol is that it uses the response to a test dose to monitor and correct for sensitivity changes that occur when a single aliquot is repeatedly used; this is the most robust measurement procedure currently available. Still, it is pointed out that the ability to accurately re-measure a sensitivity-corrected response to a given heat and dose-treatment does not necessarily guarantee that sensitivity-corrected responses to varying heat and/or dose-treatments also accurately relate to each other.

2.8 Precision

At its best, the uncertainty on a luminescence age is 6-10%; any age quoted with less uncertainty should not be trusted.[15] This is because the total uncertainty is composed of a contribution from random and systematic sources of uncertainty. Random uncertainties relate to how precise the average D_e and annual dose is known, which means that they can be reduced by performing more measurements. All other uncertainties relate to systematic effects, which originate, for example, with calibration of the equipment, nuclear data (to convert radionuclide concentrations to dose rates), dose rate correction factors (e.g. for beta attenuation), and the estimation of the time-averaged water content and the cosmic dose rate. Ultimately, sources of systematic uncertainty limit the precision that can be achieved as they cannot be reduced by performing more measurements. Although uncertainty analysis in luminescence dating has been well-formalised for many years,[16] it is often missing from dating-reports and it is of concern that users of luminescence dates may not even consider it at all. It is stressed that a luminescence date is not just a single number; in fact, it is meaningless without its uncertainty.

When a luminescence date is compared with other luminescence dates obtained at the same or another site, or when the comparison is made with independent age control, it is important that the uncertainty on the date includes the appropriate contributions. It is pointed out that random uncertainties can propagate into systematic uncertainties (such as the random uncertainty associated with the calibration measurements turning into a systematic effect when this calibration is used to determine an unknown quantity). For a suite of samples it may be possible that some sources of systematic uncertainty (such as those associated with calibration and with the time-averaged moisture content) are shared, so that they do not need to be considered when the internal consistency of the results is evaluated; if they are not, the contribution from these sources of uncertainty must be included.

2.9 Accuracy

Murray and Olley[17] have investigated the accuracy of SAR quartz ages for sediments from a wide variety of depositional environments and ages by comparing the OSL dates with independent chronologies. They concluded that the ages are accurate and that there is no evidence for systematic errors over an age range of at least 350 ka. Madsen and Murray[18] presented a similar exercise for young sediments (<1000 years); in their plot of OSL age versus independent age control, no over- or underestimation of the OSL ages is apparent, nor is there evidence for an intercept on the OSL axis. Other studies have also reported on a good agreement between quartz-based SAR OSL ages and independent age information.[19] The overall conclusion that can be drawn from these studies is that luminescence dating is, in general, a reliable and accurate chronological tool.

Nevertheless, there are also some studies that report on a discrepancy between the SAR quartz ages and independent age control. Assuming that the latter is accurate, it points at limitations in the former, at least occasionally. The main fundamental limits to the accuracy of optical dating relate to (i) resetting of the luminescence clock, (ii) post-depositional mixing and (iii) intrinsic luminescence properties of the quartz.

2.9.1 Resetting

It has been well-documented that resetting of the luminescence clock is a heterogeneous process, in which not all the grains are exposed to light for the same amount of time or with the same intensity.[20] If all grains have received sufficient exposure to light, then the sample is well-reset (or well-bleached). While this is usually assumed to be the case for aeolian sediments, it can be questioned for samples that have been transported in different ways (such as by water). In case of incomplete resetting, the grains will be deposited with varying residual doses. If the D_e is then determined using an aliquot composed of many grains, this value will be overestimated, and hence also the age. However, through using single-aliquot protocols it has become possible to investigate the completeness of resetting explicitly by measuring the D_e distribution in a sample using many small aliquots, which are composed of only a few grains, or even single grains. For an incompletely reset sample, an asymmetric D_e distribution is obtained that is skewed to higher values. The lowest D_e-values of this distribution are more likely to reflect the true burial age. But if incomplete resetting (partial bleaching) is detected, it is still possible that even the lowest values of the dose distribution are overestimates; the test for accuracy then relies on comparison with solid independent age control. The effects of incomplete resetting become less important with the age of the samples (because of dilution of

---

15 Murray – Olley 2002; Vandenberghe 2004.

16 Aitken – Alldred 1972; Aitken 1976.

17 2002.

18 2009.

19 see e.g. Ballarini *et al.* 2003; Goble *et al.* 2004; Madson *et al.* 2004, 2005; Murray *et al.* 2008; Vandenberghe *et al.* 2004; 2009.

20 Olley *et al.* 1998; Vandenberghe *et al.* 2007.

the residual dose). Jain *et al.*[21] conclude that incomplete resetting is unlikely to affect the accuracy of the age determination for samples older than about 1000 years; they also point out that modern analogues are not necessarily representative for the bleaching history of a sample. Finally, it is suggested that the size of the aliquots should be made sufficiently small (ultimately, a single grain), for the resulting D_e distributions to be useful. While this can be easily accomplished for grains from the sand fraction (>63µm), no practical approach has been developed yet for fine silt-sized (4-11µm) grains.

2.9.2 Post-depositional mixing

Plants, animals and humans can disturb sediments in such a way that layers of different age are mixed. This is especially true for sandy deposits, which are highly prone to disturbances by fauna and flora – something that may be very difficult to recognise macroscopically in the field. The relevance of post-depositional mixing to luminescence dating has only been explored relatively recently, but it has been convincingly demonstrated that the process can significantly affect OSL ages.[22] It is worth noting, however, that even though it may not always be possible to isolate the true burial age from a mixed deposit, luminescence analysis of disturbed layers may still be useful to improve the understanding of the chronology of a sequence (e.g. if a mixed layer is intercalated between pristine sediments) and/or of mixing processes in general.

2.9.3 Intrinsic luminescence properties

The quartz OSL signal is composed of different components, each exhibiting quite different characteristics.[23] The components are called *fast*, *medium* and *slow* according to how fast they are depleted by exposure to light. So far, only the fast component has convincingly shown to be useful for D_e determination, and it is for this component that the SAR protocol was developed. The relative contribution of each of these components to the OSL signal may differ from sample to sample. Some quartz samples are not dominated by the fast component, and this may upset the accuracy of the D_e determination, or not allow determination at all. In many cases, the three procedural checks (recycling, recuperation, and dose recovery) of the SAR protocol are able to identify a problem related to signal composition. Additionally, the OSL signal can be measured through linear modulation of the stimulation power (linearly modulated OSL or LMOSL), which allows an easier component-analysis of the OSL signal.[24] Difficulties with respect to signal composition can often be overcome, or at least minimised, by adjusting the SAR measurement parameters[25] and/or the way the signal is processed.[26] Preferentially isolating the fast component requires, however, that it is present with a useful intensity. In some samples, this is not the case; alternative luminescence dating methods are then to be applied, or new ones developed.

As indicated earlier (Section 2.2), saturation of the luminescence signal limits the age range over which luminescence dating is applicable. At least in principle, signal saturation limits the precision of the age determination: a minimum age is obtained, with an asymmetric uncertainty range that extends to infinity. In practise, several authors have observed that quartz OSL signals may also underestimate the age as they approach saturation,[27] or when they intersect an additional linear component in the high-dose region of the growth curve.[28] The cause for this inaccuracy is not understood at present. Although not all samples seem to be affected,[29] results obtained from the high-dose region of the growth curve are to be considered with caution.

Another type of complexity arises if, despite repeated physical and chemical treatments, no pure quartz extract can be obtained. The purity of the quartz is usually evaluated through stimulation with infrared (IR) light, in which a significant sensitivity to IR stimulation is then attributed to the presence of feldspar.[30] Methods have been devised to deal with "dirty" samples, such as pulsing the stimulating light source[31] or performing an IR cleanout prior to stimulation with the blue diodes.[32] These approaches aim at minimising the unwanted contribution to the quartz OSL signal from feldspar; unfortunately, they do not necessarily remove it entirely, so it is prudent to test the purified signal for anomalous fading.

---

[21] 2004.

[22] Bateman *et al.* 2003; 2007; Vandenberghe *et al.* 2009.

[23] Jain *et al.* 2003.

[24] Bulur 1996; Jain *et al.* 2003.

[25] see e.g. Choi *et al.* 2003a; Murray – Wintle 2003.

[26] see e.g. Choi *et al.* 2003b; Ballarini *et al.* 2007.

[27] Murray – Funder 2003; Buylaert *et al.* 2007; Murray *et al.* 2007.

[28] Timar *et al.* 2010.

[29] Murray *et al.* 2008.

[30] Duller 2003; Vandenberghe 2004.

[31] Denby *et al.* 2006; Thomsen *et al.* 2008a.

[32] Banerjee *et al.* 2001; Wallinga *et al.* 2002.

2.10 The importance of luminescence dating

Luminescence dating allows the direct determination of depositional ages for sediments from a wide range of depositional environments. It does not use associated material, but is applied to the main mineral constituents of the sediment itself (i.e. quartz and feldspar). As quartz and feldspar are the most abundant minerals in the sedimentary environment, and both silt-and sand-sized grains can be used, there is rarely a lack of datable material. The method is applicable over an age range of a few years to ~100 ka, and it has the potential to go even further back in time; as such, it has quite a wide dynamic range.

Luminescence dating has undergone major methodological developments in the past ten years. This resulted in a significant improvement in the achievable precision and accuracy, as a result of which the technique is now considered as one of the most important chronometers in the study of the Young Quaternary. As any method, luminescence dating has its limitations with respect to precision and accuracy; unlike many other dating techniques, however, the uncertainty budget is formalised reasonably well, and internal tests are available to assess whether the accuracy is compromised.

3. Application of OSL dating to fluvial deposits in NE Syria

3.1 Sites and sampling

Over the past few years, fluvial exposures have been studied near a number of archaeological settlement mounds (tells) in the Upper Khabur Basin (fig. 6). The sections HAM III (near Tell Hamidi)[33] and GPS (near Tell Brak)[34] consist of deposits by the Wadi Jaghjagh, while the sections FAR and LEI expose sediments deposited by the Wadi Jarrah near Tell Farsuk Kabir and Tell Leilan, respectively. Detailed descriptions of the investigated sections, as well as of their geographical and archaeological setting, are to be found elsewhere in this volume and are not elaborated upon here.

The samples for luminescence dating were collected by Katleen Deckers at night, in a light tight shelter. The profiles were first cleaned by scraping off a few centimetres of sediment. The samples were then collected by either hammering tubes in, or by scooping material into a black bag. Material surrounding the OSL samples was collected for annual dose determination. After sample collection, the stratigraphic position of the samples was verified under lightened conditions. Table 2 lists the samples that have been investigated in this work.

3.2 Independent age information

The fluvial sequences generally contain little or no material suitable for dating by alternative methods. Despite this, for the sequences near to Tell Hamidi (section HAM III), Tell Leilan (section LEI VI)[35] and Tell Farsuk Kabir (section FAR) some independent age information is available through TL-screening of sherds and ^{14}C dating of organic matter.[36] The radiocarbon ages were calibrated using the OxCal program (version 3.10)[37] and the IntCal04 calibration curve.[38] To facilitate comparison with the results from optical dating, all the calibrated ages are expressed as the 68.2% probability age range in thousands of years before AD 2000.

At this point, attention is drawn to the fact that different dating methods also date different events: OSL dating determines the time that has elapsed since mineral grains were last exposed to sunlight, TL-screening estimates when a sherd was heated for the last time, and a radiocarbon date relates to fixation of atmospheric carbon. When comparing the results obtained using the different methods, it should be noted that these events were not necessarily synchronous.

3.2.1 Section FAR

The deposits from section FAR contained sherds, of which thermoluminescence (TL) screening was applied to two of them. The first dated sherd was found at the same level as sample GLL-040104 and yielded an age of 3.2±1.5 ka. The second sherd was found at a level just above that of sample GLL-040107 and was dated at 2.2±0.9 ka. At a depth of 3.6 m below the surface level, a layer of possibly re-deposited soil sediments is intercalated between samples GLL-040104 and –05. AMS ^{14}C dating of humic acids extracted from this organic-rich sediment layer (sample KIA24912) yielded an age of 5.52-5.64 ka.

---

[33] see Deckers, this volume chapter 7.

[34] This is profile GPS 12 Cut C (labeling from T. Wilkinson) and BKS92 (of the Brak survey labeling).

[35] see Deckers, this volume chapter 6.

[36] Deckers - Vandenberghe 2007.

[37] Bronk - Ramsey 1995; 2001.

[38] Reimer *et al.* 2004.

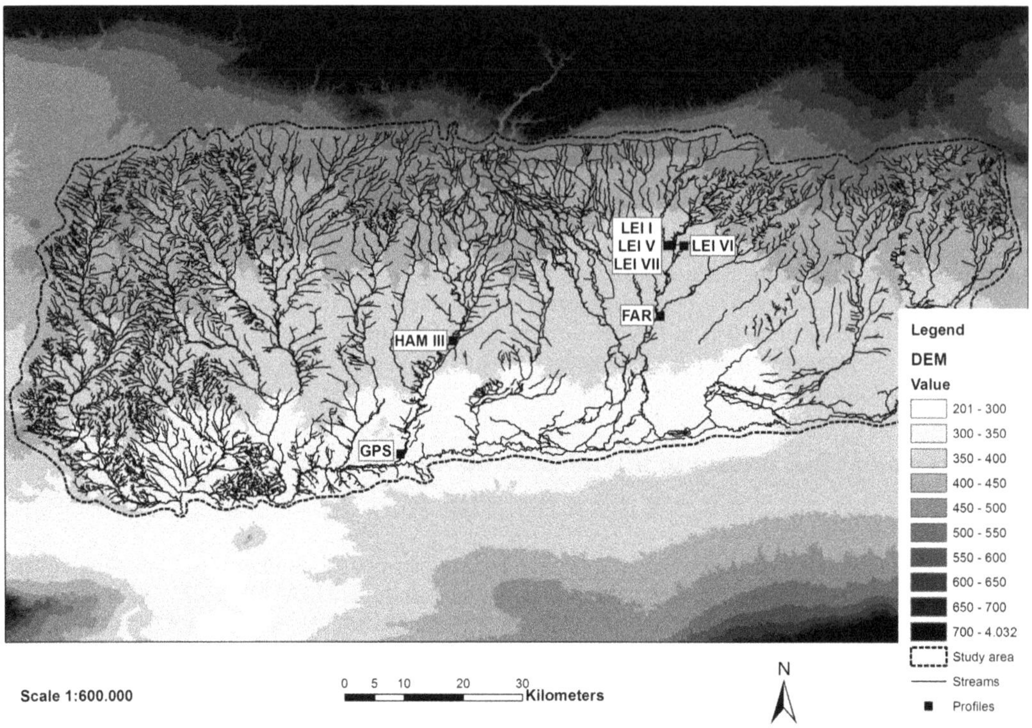

Fig. 6: Map of the Upper Khabur basin showing the location of the investigated sequences.

Table 2: Radionuclide activities, calculated dose rates, equivalent doses (D_e), calculated optical ages, and random (σ_r), systematic (σ_{sys}) and total uncertainties (σ_{tot}). The uncertainties mentioned with the D_e and dosimetry data are random; all uncertainties represent one sigma. The samples have been arranged according to their stratigraphic position (from base to top) in each sequence. The D_e for samples GLL-060111 and –12 was not measured (n.m.); see text for details. *: The thickness of the overburden was not precisely measured, but comprises an additional few metres of sediment.

Field code	GLL code	Depth (m)	Quartz fraction (μm)	^{238}U (Bq kg^{-1})	^{226}Ra (Bq kg^{-1})	^{210}Pb (Bq kg^{-1})	^{232}Th (Bq kg^{-1})	^{40}K (Bq kg^{-1})	Dose rate (Gy ka^{-1})	D_e (Gy)	Age (ka)	σ_r (%)	σ_{sys} (%)	σ_{tot} (%)	σ_{tot} (ka)
FAR OSL 1	040103	4.00	4-11	33±2	37±1	32±2	32.5±0.4	379±3	2.39±0.04	8.0±0.2	3.4	2.53	7.01	7.45	0.3
FAR OSL 2	040104	3.75	4-11	24±3	39±1	30±2	32.8±0.5	393±4	2.35±0.03	7.7±0.1	3.3	2.34	6.93	7.32	0.2
FAR OSL 3	040105	3.55	4-11	26±2	40±2	30±3	33.8±0.4	395±4	2.40±0.04	6.1±0.9	2.6	15.06	6.95	16.59	0.4
FAR OSL 4	040112	2.70	4-11	26±1	41±2	32±2	33.8±0.5	397±4	2.45±0.03	7.2±0.1	2.9	1.50	6.93	7.09	0.2
FAR OSL 5	040107	1.90	4-11	20±2	41±3	25±2	20±2	361±3	2.19±0.03	5.2±0.3	2.4	6.87	6.85	9.70	0.2
FAR OSL 6	040113	1.50	4-11	24±2	41±3	28±3	30±1	355±5	2.23±0.04	5.5±0.1	2.5	2.56	6.86	7.33	0.2
HAM III OSL 1	040101	3.40	90-125	7±1	18±1	11±1	8.2±0.2	60±1	0.56±0.01	3.0±0.1	5.3	3.89	6.45	7.53	0.4
HAM III OSL 2	040102	3.12	90-125	14±1	17.6±0.3	14±1	15.6±0.3	152±2	0.94±0.01	4.5±0.1	4.7	2.22	6.40	6.77	0.3
HAM III OSL 2	040102	3.12	4-11	14±1	17.6±0.3	14±1	15.6±0.3	152±2	1.12±0.2	4.7±0.1	4.2	2.25	6.83	7.19	0.3
HAM III OSL 4	040106	2.70	90-125	10±1	29±1	15±1	15.4±0.2	149±3	0.93±0.02	5.2±0.2	5.6	5.00	6.38	8.10	0.5
HAM III OSL 5	040109	1.60	4-11	20±1	31±2	20±2	23.8±0.3	275±4	1.75±0.03	8.1±0.1	4.7	1.96	6.79	7.06	0.3
HAM III OSL 6	040110	0.60	4-11	22±2	27±3	23±2	25.9±0.5	295±4	1.92±0.03	5.8±0.1	3.0	2.19	6.76	7.11	0.2
GPS OSL 4	060101	3.80	4-11	19±2	22±2	19±2	17.4±0.3	213±3	1.40±0.04	8.9±0.3	6.4	4.14	6.82	7.98	0.5
GPS OSL 2	060102	3.70	4-11	13±1	21±1	19±2	14.8±0.3	149±3	1.16±0.04	3.7±0.1	3.2	3.77	6.87	7.84	0.3
GPS OSL 3	060103	3.60	4-11	17±2	30±2	24±3	19.6±0.4	198±3	1.47±0.04	4.2±0.1	2.9	3.87	6.95	7.95	0.2
LEI I. OSL 3	060105	2.70	4-11	19±2	30±3	26±3	25.7±0.4	277±4	1.85±0.04	5.1±0.2	2.8	4.08	6.87	7.99	0.2
LEI VI. OSL 1	060107	3.95	4-11	15±1	14±1	17±1	9.7±0.3	105±2	0.92±0.02	43±1	47	3.17	6.88	7.58	4
LEI VI. OSL 2	060108	3.30	4-11	20±3	27±2	25±1	22.3±0.4	256±4	1.70±0.03	19.0±0.7	11.1	4.24	6.87	8.07	0.9
LEI V. OSL 1	060109	2.60	4-11	31±3	34±2	31±2	29.9±0.4	386±5	2.36±0.04	77±2	33	3.06	6.88	7.53	2
LEI V. OSL 2	060110	2.25	4-11	18±2	27±1	29±1	18.3±0.3	200±3	1.56±0.02	70±1	45	1.73	6.91	7.12	3
LEI VII. OSL 1	060111	>2.50*	4-11	24±2	19±2	14±2	21.0±0.4	249±3	1.54±0.03	n. m.	-	-	-	-	-
LEI VII. OSL 2	060112	>2.10*	4-11	22±3	24±2	19±2	21.6±0.3	225±3	1.55±0.04	n. m.	-	-	-	-	-

3.2.2 Section HAM III

At a depth of roughly 6.5 m, in section HAM III, a subsurface organic layer was observed in a bore-core. It contained charred plant remains that were AMS ^{14}C-dated to 5.37-5.51 ka (sample KIA24908). This organic layer can be correlated with an organic-rich level near the base of section HAM III, below the location of the lowermost OSL sample (GLL-040101). Additionally, a shell of a bivalve mollusc from a level underlying the organic-rich layer in section HAM III was AMS ^{14}C-dated to 6.49-6.65 ka.

3.2.3 Section LEI VI

For section LEI VI, a shell from a freshwater mussel (*Unio*) was found at exactly the same level as sample GLL-060108, but on the other side of the profile wall. The shell yielded an AMS ^{14}C age of 5.81-5.96 ka (sample KIA30899). Furthermore, using infrared stimulated luminescence signals (IRSL) from polymineral fine grains, Hilgers[39] obtained an age of 4.04±0.4 ka for a sediment sample that was collected seventy centimetres above sample GLL-060108.

3.3 Sample preparation and analytical facilities

The strategy adopted for sample preparation followed from experience with a first batch of samples (sections FAR and HAM III). For these, it was observed that only three out of twelve samples contained sufficient coarse quartz grains; the majority of the samples required fine-grain dating. As it is desirable to streamline laboratory procedures, samples from following batches (sections LEI and GPS) were immediately processed so as to obtain the fine grains.

Coarse (90-125µm) quartz grains were extracted from three samples (GLL–040101, –02 and –06) using conventional sample preparation procedures (i.e. HCl, H_2O_2, sieving, heavy liquids, HF).[40] Depending on the availability of material, aliquots were prepared by spreading the quartz grains out on the inner two millimetres or seven millimetres of 9.7 mm diameter stainless steel discs. For the remainder of the samples, polymineral fine (4-11µm) grains were isolated in the usual manner,[41] from which quartz grains were then extracted by etching for either five days using H_2SiF_6, or seven minutes using 40% HF. Aliquots were prepared by settling a 2 mg/ml suspension on aluminium discs. Table 2 indicates the grain-size fraction that was used for each sample; for sample GLL-040102, both fine (4-11µm) and coarse (90-125µm) quartz grains were extracted.

The purity of all quartz extracts was confirmed by the absence of a significant IRSL response at 60°C to a large regenerative β-dose. The sensitivity to infrared stimulation was defined as significant if the resulting IRSL signal amounted to more than 10% of the corresponding OSL signal[42] or if the IR depletion ratio deviated more than 10% from unity.[43]

All OSL measurements were made using automated Risø readers equipped with blue (470 ± 30 nm) diodes; stimulation with infrared light was through IR diodes (875 nm) or an IR laserdiode (830 nm). Luminescence emissions were detected through a 7.5 mm thick Hoya U-340 filter. Details on the measurement apparatus can be found in Bøtter-Jensen *et al.*[44]

Low-level high-resolution gamma-ray spectrometry in the laboratory was used for the determination of the dose rate. Sediment collected from the surroundings of the OSL tubes was oven-dried (at 110°C until constant weight), pulverised and homogenised, and stored in a plastic container for at least one month before being counted on top of the detector. Details on the instrumentation, spectrum evaluation and radionuclide concentration calculation can be found in Vandenberghe.[45]

3.4 Luminescence characteristics and equivalent dose determination

The equivalent dose (D_e) was determined using the single-aliquot regenerative-dose (SAR) protocol (see Section 2.7).[46] Optical stimulation with the blue diodes was for forty seconds. The initial 0.32 s of the decay curve minus a background evaluated from the last four seconds was used in the calculations. The SAR protocol that was used for D_e determination, involved a preheat of ten seconds at 220°C and a cutheat

---

[39] This volume chapter 11.

[40] see Vandenberghe *et al.* 2004.

[41] Lang *et al.* 1996; Frechen *et al.* 1996.

[42] Vandenberghe 2004.

[43] Duller 2003.

[44] 2003.

[45] 2004.

[46] Murray – Wintle 2000; 2003.

to 160°C; samples GLL-060111 and –12 were investigated using a slightly higher preheat of 10 s at 240°C. After the measurement of the response to the test dose, a high-temperature clean-out was performed by stimulation with the blue diodes for forty seconds at 280°C. For each aliquot that was measured, a complete growth curve was constructed based on the measurement of the response to five regenerative doses (up to at least ~2.5 times the estimated D_e). The response to a zero dose was measured to see whether the dose response passes through the origin (recuperation), after which a second measurement of the response to the first regenerative dose was carried out to verify whether the sensitivity correction was working well (recycling). The response to the highest regenerative dose was also measured twice, the second time to check the sensitivity to stimulation with infrared light before stimulation with the blue diodes. The purpose of this treatment was to allow identification of aliquots with a significant feldspar contamination. A D_e value was accepted if the recuperation and IRSL/OSL ratio did not exceed a threshold of 5% and 10%, respectively, and if both the recycling ratio and the depletion ratio were within 10% from unity.

The inset to Figure 7a shows the natural and regenerated OSL signals for an aliquot of fine-grained quartz from one of the samples (sample GLL-040102); both are identical in shape and are characteristic for signals that are dominated by the fast component. These are the typical OSL signals observed for the quartz extracts. All signals could be well-distinguished from the background level. Sample GLL-040105 emitted natural signals of less than 100 counts s^{-1}; while this was still detectable, it lead to results that are not as precise as for the other samples.

A representative SAR growth curve is shown in Figure 7a. The growth of luminescence could be represented well by a single saturating exponential function, or the sum of a single saturating exponential and a linear component. Figure 7a illustrates the generally good behaviour of the samples in the SAR protocol, with recycling ratios indistinguishable from unity and recuperated signals amounting to no more than ~1% of the sensitivity-corrected natural signal. Figure 7b plots the response to the test dose as a function of SAR measurement cycle (the signals have been normalised to the test-dose signal appropriate to the natural OSL signal). It illustrates that sensitivity changes do occur, and that the biggest change occurs from the first to the second measurement cycle. In the given example, the sensitivity increases by a factor of two; not all aliquots exhibited this amount of sensitivity change.

The performance of the applied SAR procedure was evaluated through a dose recovery test. Natural aliquots were first bleached twice for 250 seconds using the blue diodes at room temperature; the two bleaching treatments were separated by a 10000 second pause. The aliquots were then given a known laboratory beta dose chosen to approximate the natural burial dose, and finally measured using the SAR protocol. The dose recovery test was performed for at least three aliquots of each sample and the results are summarised in Figure 8. It can be seen that, for most samples, the given dose can be recovered to within 5% (fig. 8a); recycling ratios fall within 10% from unity and recuperation is low (fig. 8b). A similar pattern of sensitivity change was observed as for the measurements of D_e (see higher); it varied both between aliquots and samples and the change could amount to a factor of ~2 from the first to the second measurement cycle. Three samples show a poor dose recovery (GLL-060110, –11 and –12) with sample GLL-060112, in particular, overestimating the given dose by a factor of two. Interestingly, preliminary measurements showed the D_e in this sample to be of the order of a few hundred Gy (inset to Figure 8a), which would correspond to an age of more than 100 ka. As this is beyond the age range of interest here, this sample was not considered further. Sample GLL-060111 was rejected from further analysis for the same reason.

It can be noted that a poor dose recovery can be obtained even in case of a good recycling ratio (and vice versa). This is because the recycling ratio only monitors the accuracy of the sensitivity correction throughout successive SAR cycles; it gives no information as to whether the sensitivity measured in the first cycle accurately reflects the efficiency by which the first dose, or the natural dose, created luminescence.

The results from the dose recovery test indicate that, for the majority of the samples, the SAR protocol is able to measure a laboratory dose given prior to any heat treatment both accurately and precisely. Using this protocol, in between 8 and 37 replicate measurements of the D_e were made. The average D_e values (±1 standard error) are summarised in Table 2.

3.5 Dosimetry

The specific radionuclide activities obtained using low-level high-resolution gamma-ray spectrometry in the laboratory are summarised in Table 2. These were converted to annual doses using factors derived from the nuclear energy releases tabulated by Adamiec and Aitken.[47] For the coarse quartz grains, a factor of 0.9 (±5% relative uncertainty) was adopted to correct the external beta dose for the effects of attenuation and etching.[48]

---

[47] 1998.

[48] Mejdahl 1979.

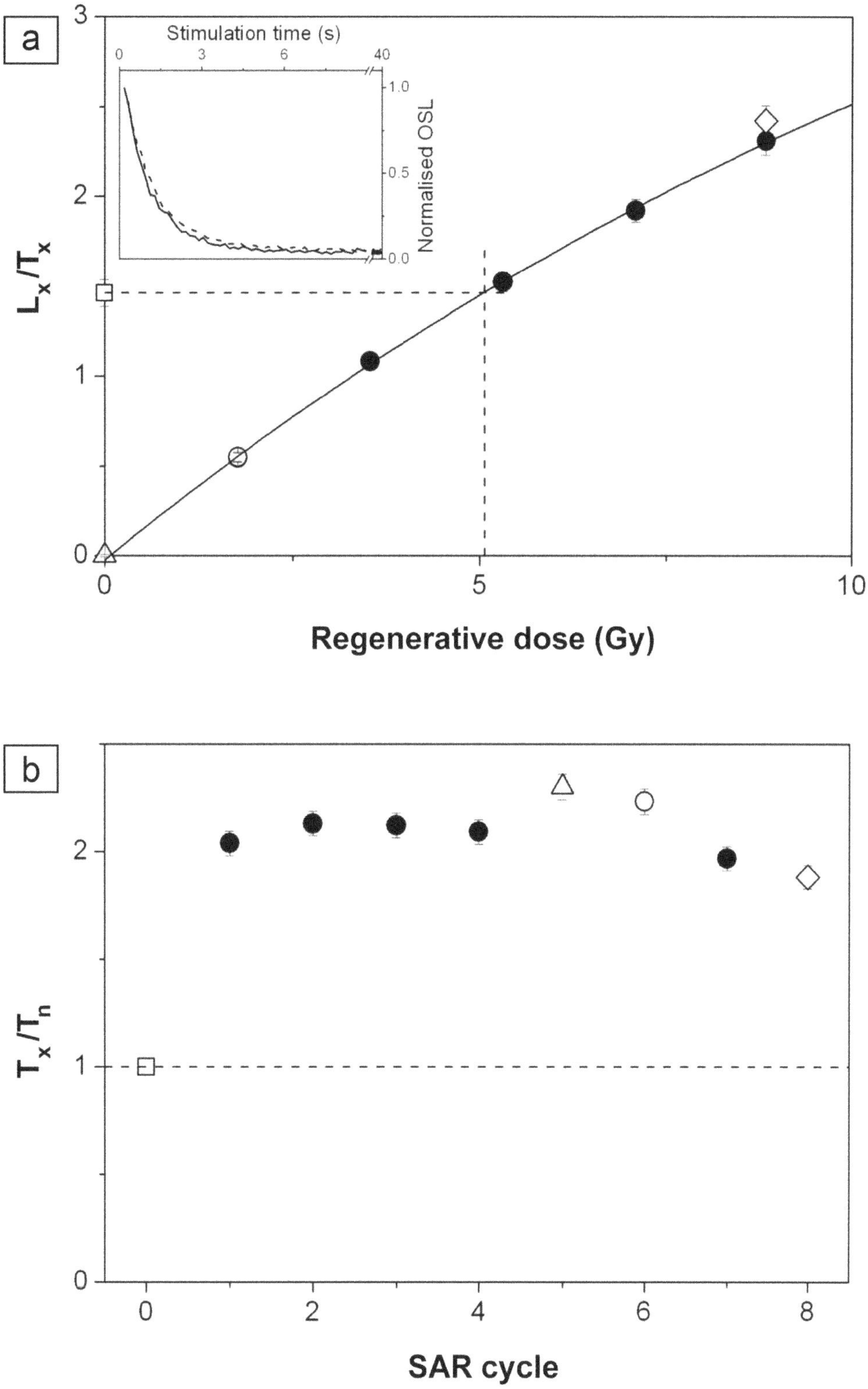

Fig. 7 a: Representative SAR growth curve obtained for an aliquot of fine-grained (4-11 μm) quartz (sample GLL-040102). Recycling and recuperation points are shown as an open circle and open square, respectively. The open diamond is the repeat measurement of the highest regenerative dose point, in which the OSL response to the regenerative dose was measured after stimulation with IR. The solid line is the fit to the data of a single saturating exponential function. The inset shows the natural (solid line) and regenerated (dashed line) OSL decay curves (normalised to the number of counts observed in the first 0.16 s of stimulation); b. Change in OSL response to the test dose (T_x) throughout the measurement of the growth curve shown in a. The data have been normalised to the test-dose response that was measured after the natural OSL signal (T_n).

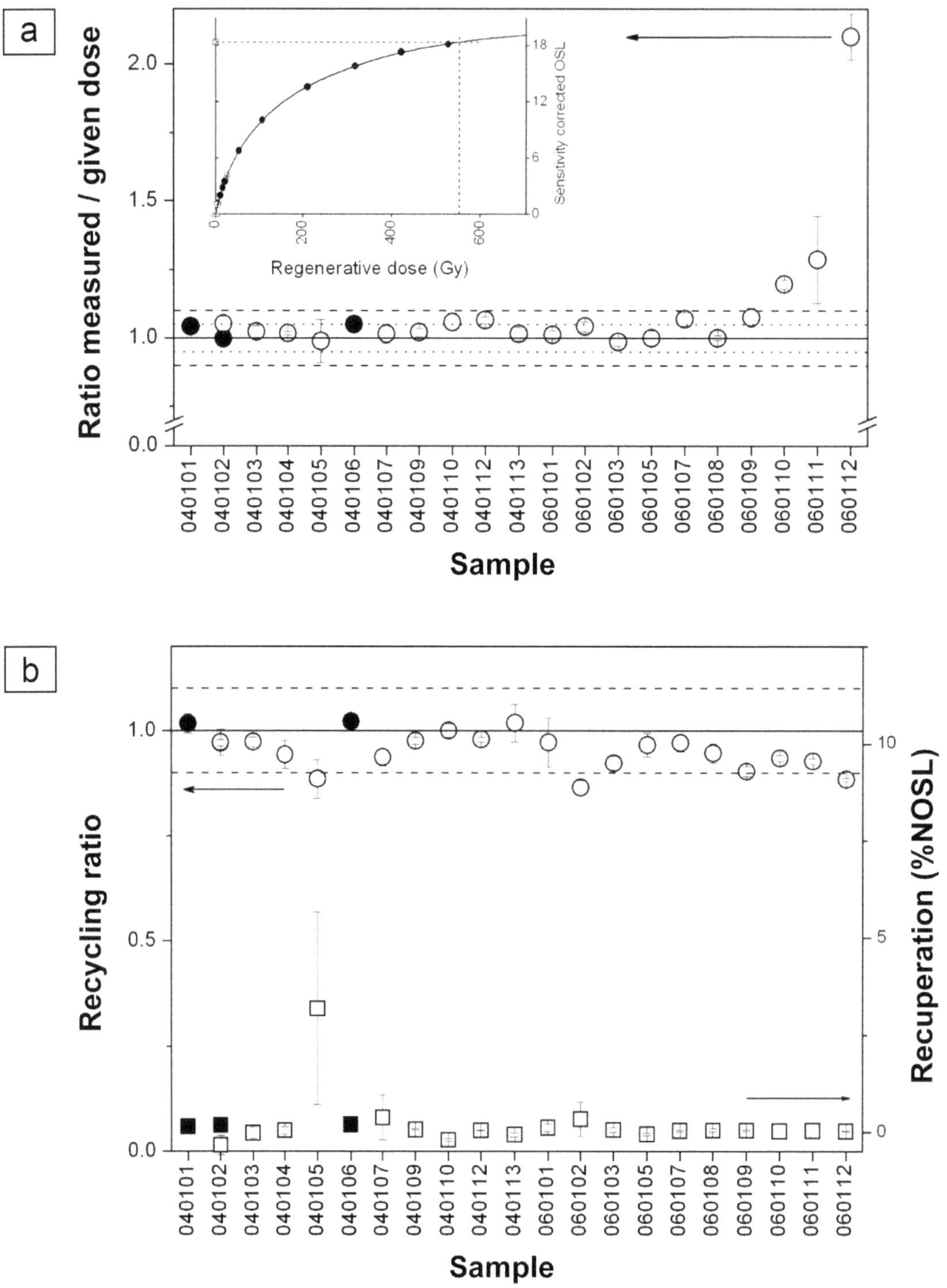

Fig. 8: Summary of dose recovery data: a. Ratio of measured to given dose. Each data point represents the average (±1 standard error) of at least three replicate measurements. Open and solid circles refer to data obtained using silt- and sand-sized quartz grains, respectively. The solid line (eye guide) represents a measured-to-given dose ratio equal to unity. The dotted and dashed lines bracket, respectively, a 5% and 10% deviation of this ratio from unity. The inset shows the sensitivity-corrected growth curve for an aliquot of sample GLL-060112; b. Corresponding values for the recycling ratio (circles) and recuperation (squares; expressed as percentage of the sensitivity corrected response to the given dose).

A value for the internal dose rate of 0.013±0.003 Gy ka^{-1} was assumed, based on Vandenberghe *et al.*[49] The time-averaged moisture content was estimated at 27±3%; this estimate is based on a value for the saturation level that falls within the range of values typically found for soil and the fact that the samples were near saturation over the entire burial period. The contribution from cosmic radiation was calculated using Prescott and Hutton.[50] For the fine quartz grains, an alpha efficiency factor of 0.04±0.01 was used;[51] the contribution from Rb was calculated assuming a K:Rb ratio of 200:1.[52]

3.6 Optical ages and discussion

Table 2 summarises the information relevant to the age and uncertainty calculation, which followed the system formalised by Aitken and Alldred[53] and Aitken.[54] It can be seen that, for most samples, the systematic uncertainty is dominant in the overall uncertainty on the age, which typically amounts to ~7-8% (1). The comparatively poor random uncertainty (~15%) for sample GLL-040105 is due to the dim signals emitted by the quartz (see Section 3.4).

No age was calculated for samples GLL-060111 and -12, owing to the poor dose recovery that was achieved. Combining the dosimetric information quoted in Table 2 with initial D_e estimates for these samples, would yield ages in excess of 200 ka. Although these age estimates are expected to be inaccurate, they do indicate that sediment of quite a considerable age can be located at a relatively shallow level below the subsurface. For the other sections, most of the dates are consistent with the stratigraphic position of the samples; only for section LEI V, is there an apparent age inversion. It can be noted that a poor dose recovery was achieved for sample GLL-060110, with the measured dose overestimating the given dose by about 20%. This could account for the apparent inconsistency. Still, the dataset is too limited (only two dates) to allow for a meaningful discussion with respect to its internal consistency and accuracy; this illustrates the importance of collecting and dating more samples per site. It is less speculative to conclude that the optical dates obtained for section LEI V point at a Pleistocene rather than a Holocene age for the sediments.

For both the FAR and the HAM III sections, six optical ages have been obtained; some independent age information is available for these sequences as well (see Section 3.2). Therefore, the following discusses the results obtained for these sections in greater detail. Although only two optical dates were obtained for section LEI VI, they are included in the discussion; the reasons for doing so are that they (i) illustrate a strength of luminescence dating, i.e. its ability to detect erosive hiatuses, and (ii) indicate that a single date should never be used to establish a chronology for a site.

3.6.1 Section HAM III

Figure 9 plots the optical ages obtained for the HAM III section as a function of the depth of the samples below the surface level. For this section, the ages have been obtained partly using fine-grained (4-11 µm; circles) and partly using coarse-grained (90-125 µm; squares) quartz. Both datasets contain no outliers, and they are consistent with each other. It should be noted that the error bars represent the 1 sigma uncertainty, and that random (black error bars) and total (grey error bars) uncertainties are used to evaluate the consistency within and between the datasets, respectively. One sample (GLL-040102) has been dated using both fine and coarse-grained quartz. It is reassuring that the age results agree as it has been observed that, at least in some fluvial environments, the luminescence clock in the fine and the coarse quartz grains is not reset to the same extent.[55] The entire dataset is consistent with the stratigraphic position of the samples and, within the available time-resolution, there is no evidence for the presence of an erosive hiatus.

The optical ages are also consistent with the ^{14}C dates of 5.37-5.51 ka and 6.49-6.65 ka for samples that were collected near the base of the sequence. The good agreement between the OSL results and the ^{14}C ages suggests that the samples received sufficient light exposure for resetting of the OSL clock, as was already indicated by the consistency of the OSL dataset.

3.6.2 Section FAR

The age-depth plot for section FAR is shown in Figure 10. For this section, all ages are based on OSL signals from fine-grained quartz. Within analytical uncertainty (1 random uncertainty; black error bars),

---

49 2008.

50 1994.

51 Rees-Jones – Tite 1997.

52 Aitken 1985.

53 1972.

54 1976.

55 Wallinga 2002; Truelsen – Wallinga 2003; Alexanderson 2007; Vandenberghe *et al.* 2007.

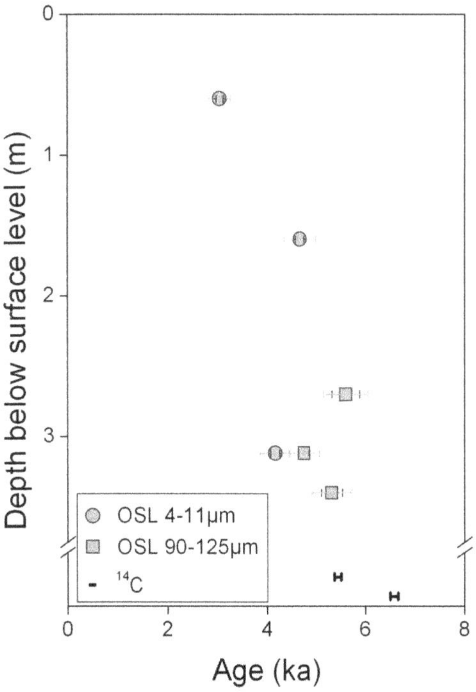

Fig. 9: Plot of ages obtained for section HAM III as a function of the depth of the samples below the surface level. The random and total uncertainties associated with the optical ages are shown by the black and grey error bars (1 sigma), respectively. The calibrated radiocarbon ages are represented by an age range.

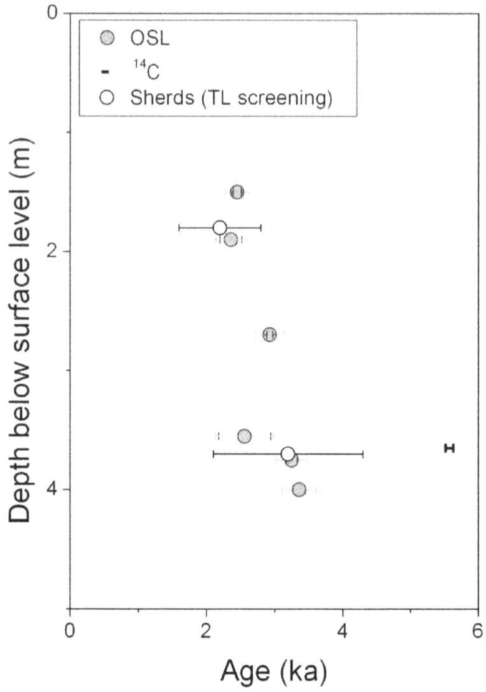

Fig. 10: Plot of ages obtained for section FAR as a function of the depth of the samples below the surface level. The random and total uncertainties associated with the optical ages (filled circles) are shown by the black and grey error bars (1 sigma), respectively. The calibrated radiocarbon age is represented by an age range; the results from TL-screening of sherds by open circles.

the optical dates are consistent with the stratigraphic position of the samples and they indicate quasi-continuous sediment deposition between ~3.4 ka and ~2.4 ka.

The dates are consistent with the results of TL screening of sherds that were found within the deposits (2.2±0.9 ka and 3.2±1.5 ka), but they conflict with the radiocarbon age of 5.52-5.64 ka obtained on humic acids extracted from a possible redeposited soil horizon. As the OSL ages have been obtained using quartz OSL signals that are dominated by a fast component (and hence are thermally stable) and a measurement protocol that passed all procedural checks, it is unlikely that they underestimate the true burial age. Therefore, the ^{14}C date must be considered as too old. This does not necessarily imply that the optical ages are not affected by incomplete resetting, although their stratigraphic consistency suggests that this is not a significant source of error.

3.6.3 Section LEI VI

Figure 11 shows the age-depth plot for the LEI VI section. Two ages of 11.1±0.9 ka and 47±4 ka have been obtained, and they are in the correct stratigraphic order.

Included in Figure 11 is the luminescence date of 4.04±0.4 ka that was independently obtained by Hilgers[56] on a sediment sample that was collected at a higher level in the sequence; the date is based on IRSL signals from polymineral fine grains that were measured using a SAR protocol. As exposure to infrared light stimulates the luminescence from feldspar, these signals are prone to anomalous fading, which leads to age underestimation. On the other hand, IRSL signals are not reset as fast and/or as completely as the fast component of the quartz OSL signal;[57] this implies that age overestimation due to incomplete

[56] chapter 11 this volume.

[57] see e.g. Vandenberghe *et al.* 2007; Thomsen *et al.* 2008b.

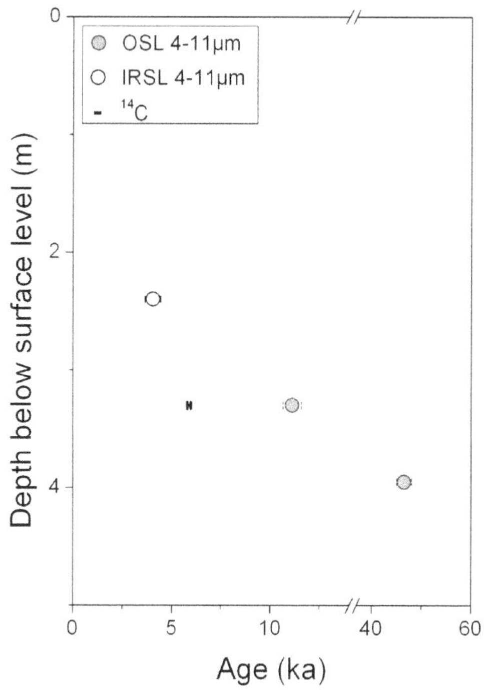

Fig. 11: Plot of ages obtained for section LEI VI as a function of the depth of the samples below the surface level. The random and total uncertainties associated with the optical ages (filled circles) are shown by the black and grey error bars (1 sigma), respectively. The calibrated radiocarbon age is represented by an age range; the IRSL age (Hilgers, this volume chapter 11) by an open circle.

resetting is of greater concern when IRSL signals are used. Hence, the age of 4.04±0.4 ka should be considered with caution. Nevertheless, it is consistent with the dataset obtained on the fine quartz grains.

The quartz OSL age of 11.1±0.9 ka, on the other hand, is not consistent with the radiocarbon date of 6.49-6.65 ka for a shell that was found at the same level (but on the other side of the profile wall). It has been pointed out that incomplete resetting is unlikely to lead to significant age overestimation for samples that are more than a few thousand years old (Section 2.9.1.). Also, the results obtained for the HAM III and FAR sections do not suggest that, in this type of environment, incompletely resetting is a significant source of inaccuracy. It is beyond the scope of this paper to discuss why the ^{14}C date may be accurate or not; it is well known that ^{14}C dating of organic materials extracted from fluvial environments is not without its difficulties.[58] In any case, the internal consistency of a dataset that is made up of a single ^{14}C date cannot be evaluated, and, therefore, this date provides no robust means to evaluate the accuracy of the OSL ages. Of course, the OSL dataset is also too limited to ensure that the age of 11.1±0.9 ka is not an outlier. This illustrates the importance of obtaining more ages for a single section, regardless of the dating method being employed.

Perhaps of more direct importance is the age jump of at least 36 ka (i.e. from ~11 to ~47 ka) over a vertical distance of less than 1 m. It is well known that fluvial archives are often incomplete, and that the hiatuses are not always recognisable in the field. The optical dates obtained for section LEI VI definitely point to this direction, as incomplete resetting can hardly be invoked to question the Pleistocene age for the lowermost sample. Hilgers[59] also observed a significant jump in age for another section along the Wadi Jarrah near Tell Leilan (profile LEI IV); there the ages jump from ~30 ka to ~2.6 ka over a vertical distance of about fifty centimetres. While these ages might be affected by anomalous fading and/or incomplete resetting, it is most unlikely that the jump in age originates with either one of these effects.

---

[58] see Howard *et al.* 2009.

[59] chapter 11 this volume.

4. Conclusions

The results obtained for the sections HAM III, FAR and LEI VI demonstrate that optical dating is a powerful tool for establishing chronologies for fluvial sediments around archaeological sites in NE Syria. Very often, it is the only method that can be applied as material suitable for dating by other methods is either lacking, or yields only indirect information on the timing of the fluvial landscape evolution.

Fluvial deposits in the study area are predominantly fine-grained sediments. Very often, they contain insufficient sand-sized quartz grains, so that optical dating must be applied to the silt-sized fraction. The use of quartz grains from this fraction is then preferred, as the quartz OSL signal is generally not affected by anomalous fading and its fast component is more readily reset in nature. Methods to detect incomplete resetting in fine-grained quartz extracts have yet to be developed, and so this remains a potential source of inaccuracy. Nevertheless, at least for the three sequences discussed in Section 3.6, it seems quite unlikely that the ages are significantly overestimated owing to incomplete resetting.

The OSL ages should always be quoted with their uncertainties, which are to be obtained through a formal analysis. Furthermore, when evaluating the internal consistency of optical dates, or when comparing them with ages obtained through alternative luminescence techniques or other dating methods, the contribution from the relevant sources of uncertainty must be included. A strong case is made for dating multiple samples from each site and section. The internal consistency of a dataset cannot be evaluated if it consists of only one or two dates. Failures do occur, and a chronology for a site should never be based on a single age estimate. This not only applies to optical dating but also to more established methods, such as radiocarbon dating. The results presented in Section 3.6 demonstrate the importance of a strategy that involves the dating of a sufficiently large number of samples. They also illustrate that (geo)archaeologists should not uncritically apply radiocarbon dating in fluvial environments.

The ability to recognise hiatuses is a strong feature of luminescence dating. Where this is not recognised in the field, the age results may challenge basic ideas on geomorphic processes and the significance of the deposits,[60] which, in turn, may lead to an improved understanding of the environmental variables around an archaeological site. At the very least, luminescence dating is a powerful tool to identify that part of the investigated record that is missing; this is of direct relevance to the (geo)archaeologist as it helps avoiding erroneous palaeoclimatic and palaeoenvironmental reconstructions.

Bibliography

ADAMIEC, G. – AITKEN, M.J.
1998 Dose-rate conversion factors: update, Ancient TL 16, 37-50.

AITKEN, M.J.
1976 Thermoluminescence age evaluation and assessment of error limits: revised system, Archaeometry 18, 233-238.
1985 Thermoluminescence Dating.

AITKEN, M.J. – ALLDRED, J.C.
1972 The assessment of error limits in thermoluminescence dating, Archaeometry 14, 257-267.

ALEXANDERSON, H.
2007 Residual OSL signals from modern Greenlandic river sediments, Geochronometria 26, 1-9.

BAILEY, R.M. – ADAMIEC, G. – RHODES, E.J.
2000 OSL properties of NaCl relative to dating and dosimetry, Radiation Measurements 32, 717-723.

BALLARINI, M. – WALLINGA, J. – MURRAY, A.S. –VAN HETEREN, S. – OOST, A.P. – BOS, A.J.J. – VAN EIJK, C.W.E.
2003 Optical dating of young coastal dunes on a decadal time scale, Quaternary Science Reviews 22, 1011-1017.

BALLARINI, M. – WALLINGA, J. – WINTLE, A.G. – BOS, A.J.J.
2007 A modified SAR protocol for optical dating of individual grains from young quartz samples, Radiation Measurements 42, 360-369.

BANERJEE, D. – MURRAY, A.S. – BØTTER-JENSEN, L. – LANG, A.
2001 Equivalent dose estimation using a single aliquot of polymineral fine grains, Radiation Measurements 33, 73-94.

[60] see e.g. Stevens *et al.* 2006; Buylaert *et al.* 2008.

BATEMAN, M.D. – FREDERICK, C.D. – JAISWAL, M.K. – SINGHVI, A.K.
2003 Investigations into the potential effects of pedoturbation on luminescence dating, Quaternary Science Reviews 22, 1169-1176.

BATEMAN, M.D. – BOULTER, C.H. – CARR, A.S. – FREDERICK, C.D. – PETER, D. – WILDER, M.
2007 Detecting post-depositional sediment disturbance in sandy deposits using optical luminescence, Quaternary Geochronology 2, 57-64.

BOWMAN, S.G.E. – HUNTLEY, D.J.
1984 A new proposal for the expression of alpha efficiency in TL dating, Ancient TL 2, 6-11.

BRONK RAMSEY, C.
1995 Radiocarbon calibration and analysis of stratigraphy: the OxCal program, Radiocarbon 37, 425-430.
2001 Development of the radiocarbon program OxCal, Radiocarbon 43, 355-363.

BULUR, E.
1996 An alternative technique for optically stimulated luminescence (OSL) experiment, Radiation Measurements 26, 701-709.

BUYLAERT, J.-P. – VANDENBERGHE, D. – MURRAY, A.S. – HUOT, S. – DE CORTE, F. – VAN DEN HAUTE, P.
2007 Luminescence dating of old (>70ka) Chinese loess: a comparison of single-aliquot OSL and IRSL techniques, Quaternary Geochronology 2, 9-14.

BUYLAERT, J.-P. – MURRAY, A.S. – HUOT, S.
2008 Optical dating of an Eemian site in Northern Russia using K-feldspar, Radiation Measurements 43, 715-720.

BØTTER-JENSEN, L. – ANDERSEN, C.E. – DULLER, G.A.T. – MURRAY, A.S.
2003 Developments in radiation, stimulation and observation facilities in luminescence measurements, Radiation Measurements 37, 535-541.

CHOI, J.H. – MURRAY, A.S. – JAIN, M. – CHEONG, C.S. – CHANG, H.W.
2003a Luminescence dating of well-sorted marine terrace sediments on the southeastern coast of Korea, Quaternary Science Reviews 22, 407-421.

CHOI, J.H. – MURRAY, A.S. – CHEONG, C.S. – HONG, D.G. – CHANG, H.W.
2003b The resolution of stratigraphic inconsistency in the luminescence ages of marine terrace sediments from Korea, Quaternary Science Reviews 22, 1201-1206.

DECKERS, K. – RIEHL, S.
2007 Fluvial environmental contexts for archaeological sites in the Upper Khabur basin (northeastern Syria), Quaternary Research 67, 337-348.

DECKERS, K. – VANDENBERGHE, D.
2007 Confronting luminescence with radiocarbon dates for fluvial deposits in the Upper Khabur basin of northeastern Syria, in: B. Ooghe – G. Verhoeven (eds.), Broadening Horizons – Multidisciplinary Approaches to Landscape Study, 50-74.

DENBY, P.M. – BØTTER-JENSEN, L. – MURRAY, A.S. – THOMSEN, K.J. – MOSKA, P.
2006 Application of pulsed OSL to the separation of the luminescence components from a mixed quartz/feldspar sample, Radiation Measurements 41, 774-779

DULLER, G.A.T.
2003 Distinguishing quartz and feldspar in single grain luminescence measurements, Radiation Measurements 37, 161-165.

FRECHEN, M. – SCHWEITZER, U. – ZANDER, A.
1996 Improvements in sample preparation for the fine grain technique, Ancient TL 14, 15-17.

GOBLE, R.J. – MASON, J.A. – LOOPE, D.B. – SWINEHART, J.B.
2004 Optical and radiocarbon ages of stacked paleosols and dune sands in the Nebraska Sand Hills, USA, Quaternary Science Reviews 23, 1173–1182.

HOWARD, A.J. – GEAREY, B.R. – HILL, T. – FLETCHER, W. – MARSHALL, P.
2009 Fluvial sediments, correlations and palaeoenvironmental reconstruction: The development of robust radiocarbon chronologies, Journal of Archaeological Science 36, 2680-2688.

HUNTLEY, D.J. – LAMOTHE, M.
2001 Ubiquity of anomalous fading in K-feldspars and the measurement and correction for it in optical dating, Canadian Journal of Earth Sciences 38, 1039-1106.

JAIN, M. – MURRAY, A.S. – BØTTER-JENSEN, L.
2003 Characterisation of blue-light stimulated luminescence components in different quartz samples: implications for dose measurements. Radiation Measurements 37, 441-449.
2004 Optically stimulated luminescence dating: how significant is incomplete light exposure in fluvial environments?, Quaternaire 15, 143-157.

KARS, R.H. – WALLINGA, J. – COHEN, K.M.
2008 A new approach towards anomalous fading correction for feldspar IRSL dating – tests on samples in field saturation, Radiation Measurements 43, 786-790.

LAMOTHE, M. – AUCLAIR, M. – HAMZAOUI, C. – HUOT, S.
2003 Towards a prediction of long-term anomalous fading of feldspar IRSL, Radiation Measurements 37, 493-498.

LANG, A. – LINDAUER, S. – KUHN, R. – WAGNER, G.A.
1996 Procedures used for optically and infrared stimulated luminescence dating of sediments, Ancient TL 14, 7-11.

LIAN, O.B. – HU, J. – HUNTELY, D.J. – HICOCK, S.R.
1995 Optical dating studies of Quaternary organic-rich ediments from southwestern British Columbia and northwestern Washington State, Canadian Journal of Earth Sciences 32, 1194-1207.

MADSEN, A.T. – MURRAY, A.S.
2009 Optically stimulated luminescence dating of young sediments: a review, Geomorphology 109, 3-16.

MADSEN, A.T. – MURRAY, A.S. – ANDERSEN, T.J. – PEJRUP, M. – BREUNING-MADSEN, H.
2005 Optically stimulated luminescence dating of young estuarine sediments: a comparison with 210Pb and 137Cs dating, Marine Geology 214, 251-268.

MASON, J.A. – SWINEHART, J.B. – GOBLE, R.J. – LOOPE, D.B.
2004 Late-Holocene dune activity linked to hydrological drought, Nebraska Sand Hills, USA, The Holocene 14, 209-217.

MEJDAHL, V.
1979 Thermoluminescence dating: beta-dose attnueation in quartz grains, Archaeometry 21, 61-72.

MURRAY, A.S. – FUNDER, S.
2003 Optically stimulated luminescence dating of a Danish Eemian coastal marine deposits: a test of accuracy, Quaternary Science Reviews 22, 1177-1183.

MURRAY, A.S. – OLLEY, J.M.
2002 Precision and accuracy in the optically stimulated luminescence dating of sedimentary quartz: a status review, Geochronometria 21, 1-16.

MURRAY, A.S. – WINTLE, A.G.
2000 Luminescence dating using an improved single-aliquot regenerative-dose protocol, Radiation Measurements 32, 57-73.

2003 The single aliquot regenerative dose protocol: potential for improvements in reliability, Radiation Measurements 37, 377-381.

MURRAY, A.S. – SVENDSEN, J.I. – MANGERUD, J. – ASTAKHOV, V.I.
2007 Quartz OSL age of an Eemian site on the Sula River, northern Russia, Quaternary Geochronology 2, 102-109.

MURRAY, A.S. – BUYLAERT, J.-P. – HENRIKSEN, M. – SVENDSEN, J.-I. – MANGERUD,J.
2008 Testing the reliability of quartz OSL ages beyond the Eemian, Radiation Measurements 43, 776-780.

OLLEY, J.M. – CAITCHEON, G.G. – MURRAY, A.S.
1998 The distribution of apparent dose as determined by optically stimulated luminescence in small aliquots of fluvial quartz: implications for dating young sediments, Quaternary Geochronology 17, 1033-1040.

PRESCOTT, J.R. – HUTTON, J.T.
1994 Cosmic ray contributions to dose rates for luminescence and ESR dating: large depths and long-term time variations, Radiation Measurements 23, 497-500.

REES-JONES, J. – TITE, M.S.
1997 Optical dating results for British archaeological sediments, Archaeometry 39, 177-187.

REIMER, P.J. – BAILLIE, M.G.L. – BARD, E. – BAYLISS, A. – BECK, J.W. – BERTRAND, C. – BLACKWELL, P.G. – BUCK, C.E. – BURR, G. – CUTLER, K.B. – DAMON, P.E. – EDWARDS, R.L. – FAIRBANKS, R.G. – FRIEDRICH, M. – GUILDERSON, T.P. – HUGHEN, K.A. – KROMER, B. – MCCORMAC, F.G. – MANNING, S. – BRONK RAMSEY, C. – REIMER, R.W. – REMMELE, S. – SOUTHON, J.R. – STUIVER, M. – TALAMO, S. – TAYLOR, F.W. – VAN DER PLICHT, J. – WEYHENMEYER, C.E.
2004 IntCal04 Terrestrial Radiocarbon Age Calibration, 0-26 Cal kyr BP, Radiocarbon 46, 1029-1058.

SUGISAKI, S. – BUYLAERT, J.-P. – MURRAY, A.S. – TSUKAMOTO, S. – NOGI, Y. –MIURA, H. – SAKAI, S. – IIJIMA, K. – SAKAMOTO, T.
2010 High resolution OSL dating back to MIS 5e in the central Sea of Okhotsk, Quaternary Geochronology, 5 (2-3), 293-298.

THOMSEN, K.J. – BØTTER-JENSEN, L. – MURRAY, A.S.
2002 Household and workplace chemicals as retrospective dosemeters, Radiation Protection Dosimetry 101, 515-518.

THOMSEN, K.J. – JAIN, M. – MURRAY, A.S. – DENBY, P.M. – ROY, N. – BØTTER-JENSEN, L.
2008a Minimizing feldspar OSL contamination in quartz UV-OSL using pulsed blue stimulation, Radiation Measurements 43, 752-757.

THOMSEN, K.J. – MURRAY, A.S. – JAIN, M. – BØTTER-JENSEN, L.
2008b Laboratory fading rates of various luminescence signals from feldspar-rich sediment extracts, Radiation Measurements 43, 1474-1486.

TIMAR, A. – VANDENBERGHE, D. – PANAIOTU, E.C. – PANAIOUT, C.G. – NECULA, C. – COSMA, C. – VAN DEN HAUTE, P.
2010 Optical dating of Romanian loess using fine-grained quartz, Quaternary Geochronology, 5 (2-3), 143-148.

TRUELSEN, J.L. – WALLINGA, J.
2003 Zeroing of the OSL signal as a function of grain size: investigating bleaching and thermal transfer for a young fluvial sample, Geochronometria 22, 1-8.

VANDENBERGHE, D.
2004 Investigation of the optically stimulated luminescence dating method for application to young geological sediments. PhD thesis, Ghent University. (Free download: https://biblio.ugent.be/record/000470756.)

VANDENBERGHE, D. – KASSE, C. – HOSSAIN, S.M. – DE CORTE, F. – VAN DEN HAUTE, P. – FUCHS, M. – MURRAY, A.S.
2004 Exploring the method of optical dating and comparison of optical and ^{14}C ages of Late Weichselian coversands in the southern Netherlands, Journal of Quaternary Science 19, 73-86.

VANDENBERGHE, D. – DERESE, C. – HOUBRECHTS, G.
2007 Residual doses in recent alluvial sediments from the Ardenne (S Belgium), Geochronometria 28, 1-8.

VANDENBERGHE, D. – DE CORTE, F. – BUYLAERT, J.-P. – KUČERA, J. – VAN DEN HAUTE, P.
2008 On the internal radioactivity in quartz, Radiation Measurements 43, 771-775.

VANDENBERGHE, D. – VANNESTE, K. – VERBEECK, K. – PAULISSEN, E. – BUYLAERT, J.-P. – DE CORTE, F. – VAN DEN HAUTE, P.
2009 Late Weichselian and Holocene earthquake events along the Geleen fault in NE Belgium: OSL age constraints, Quaternary International 199, 56-74.

WALLINGA, J.
2002 Optically stimulated luminescence dating of fluvial deposits: a review, Boreas 31, 303-322.

WALLINGA, J. – MURRAY, A.S. – BØTTER-JENSEN, L.
2002 Measurement of the dose in quartz in the presence of feldspar contamination, Radiation Protection Dosimetry 101, 367-370.

WINTLE, A.G.
1973 Anomalous fading of thermoluminescence in mineral samples, Nature 245, 143-144.

WINTLE, A.G. – MURRAY, A.S.
2006 A review of quartz optically stimulated luminescence characteristics and their relevance in single-aliquot regeneration dating protocols, Radiation Measurements 41, 369-391.

Appendix A: Some useful key texts and review papers

AITKEN, M.J.
1985 Thermoluminescence Dating.
1998 An introduction to optical dating .The Dating of Quaternary Sediments by the Use of Photon-Stimulated Luminescence.

BØTTER-JENSEN, L. – MCKEEVER, S.W.S. – WINTLE, A.G.
2003 Optically Stimulated Luminescence Dosimetry.

DULLER, G.A.T.
1996 Recent developments in luminescence dating of Quaternary sediments, Progress in Physical Geography 20, 127-145.
2004 Luminescence dating of Quaternary sediments: recent advances, Journal of Quaternary Science 19, 183-192.
2008 Luminescence dating: guidelines on using luminescence dating in archaeology. Swindon: English Heritage. (Free download: http://www.english-heritage.org.uk/upload/pdf/luminescence_dating.pdf.)

JAIN, M. – MURRAY, A.S. – BØTTER-JENSEN, L.
2004 Optically stimulated luminescence dating: how significant is incomplete light exposure in fluvial environments?, Quaternaire 15, 143-157.

MADSEN, A.T. – MURRAY, A.S.
2009 Optically stimulated luminescence dating of young sediments: a review, Geomorphology 109, 3-16.

MURRAY, A.S.
2000 Single Aliquot Protocols in Luminescence Dating., in: K.V.R. Murthy – M.D. Sastry – T.R. Joshi – L.H.H. Prasad - A.G. Page – N.G. Patel (eds.), Luminescence and its applications, 103-117.

MURRAY, A.S. – OLLEY, J.M.
2002 Precision and accuracy in the optically stimulated luminescence dating of sedimentary quartz: a status review, Geochronometria 21, 1-16.

PRESCOTT, J.R. – ROBERTSON, G.B.
1997 Sediment dating by luminescence: a review, Radiation Measurements 27, 893-922.

ROBERTS, R.G.
1997 Luminescence dating in archaeology: from origins to optical, Radiation Measurements 27, 819-892.

STOKES, S.
1999 Luminescence dating applications in geomorphological research, Geomorphology 29, 153-171.

VANDENBERGHE, D.
2004 Investigation of the optically stimulated luminescence dating method for application to young geological sediments. PhD thesis, Ghent University. (Free download: https://biblio.ugent.be/record/000470756.)

WALLINGA, J.
2002 Optically stimulated luminescence dating of fluvial deposits: a review, Boreas 31, 303-322.

WINTLE, A.G.
2008 Luminescence dating of Quaternary sediments, Boreas 37, 469-677.

11. Optically stimulated luminescence dating of sediment samples from Syria

Alexandra Hilgers

1. Introduction

Twelve sediment samples were dated by luminescence in order to provide information on the geochronological context of fluvial and eolian deposits from northeastern Syria. Three samples, referred to as TC OSL1-3, derive from the eolian profile at Brak;[1] another two, named OR OSL1 and 2, derive from the fluvial profile investigated in the lower town of Tell Mozan.[2] Samples TW OSL1-3 were taken from a profile from the Wadi Jaghjagh, somewhat south of Tell Barri.[3] Additionally, four OSL samples, LEI IV OSL1-3 and LEI VI OSL3, were studied from two fluvial profiles in the Leilan surroundings.[4]

Within this report detailed information will be presented concerning luminescence dating and the protocols involved. Aitken provides the most comprehensive account of luminescence dating in his books, both on thermoluminescence[5] and optically stimulated luminescence.[6] The basic principles and protocols of OSL dating and the recent advances in OSL dating of Quaternary sediments have recently been reviewed by Duller,[7] Bøtter-Jensen *et al.*,[8] Lian – Roberts,[9] and Preusser *et al.*[10]

2. Dose rate determination

In order to estimate the annual dose derived from the decay of lithogenic radionuclides, the concentration of uranium, thorium, and potassium in the sediment samples was measured by neutron activation analysis (NAA; Becquerel Laboratories, Canada). The results are summarised in Table 1. The dose rate (D_0) values (see tab. 2) include the contribution of cosmic-rays, which was determined for each sample as a function of sampling depth and geographic position.[11] All D_0 values were calculated assuming secular equilibrium for the U and Th decay chains. Since the alpha dose contribution has to be included in the dose rate determination of the fine-grained samples, an alpha-efficiency factor of 0.07±0.02[12] was used for the polymineralic fine-grained material.

Water present in pore voids attenuates the radiation dose of the individual grains more effectively than air. Hence, the overall dose rate is reduced as the moisture content rises within the sediment. Therefore, the water content of the sediment being dated has to be taken into account when calculating the dose rate. The water content of a sample 'as collected' can be measured and its influence on the radiation attenuation can be calculated by applying the appropriate attenuation factors. However, in terms of dose rate assessment for dating, the variation in wetness over the entire timespan of burial must be considered. This proves problematic. The uncertainty related with this parameter represents a fundamental limitation in improving the luminescence-age error limits to below a percentage of about 5%[13] as quantification of the range of variation in moisture is almost impossible. The range of variation in the water content during burial finally included in the dose rate calculation was based on the water contents 'as found' (see tab. 1).

---

1 see Deckers – Wilkinson, chapter 9 this volume.

2 see Deckers – Pustovoytov, chapter 4 this volume.

3 This profile TW 055 (labeling after T. Wilkinson) is not included in this volume - it is located at 36°44'18.52"N, 41° 7'30.91"E.

4 see Deckers, chapter 7 this volume.

5 Aitken 1985.

6 Aitken 1998.

7 2004.

8 2003.

9 2006.

10 2008.

11 according to Prescott – Hutton 1988; 1994.

12 Preusser *et al.* 2005.

13 Aitken 1998.

Table 1: Dose rate data, including uranium, thorium and potassium concentrations as determined by neutron activation analysis (all uncertainties represent the 1σ confidence interval). The range in the variation of water content during burial was based on the as found water contents.

Sample	Depth (m)	Water content (Weight-%)	U (ppm)	Th (ppm)	K (%)
TC OSL1	2.72	2±1	2.11 ±0.19	5.52 ±0.21	1.57 ±0.05
TC OSL2	2.12	2±1	2.35 ±0.22	5.56 ±0.22	1.66 ±0.05
TC OSL3	0.66	1±1	1.62 ±0.20	5.90 ±0.22	1.53 ±0.05
OR OSL 1	3.10	4±1	1.18 ±0.12	2.74 ±0.11	0.40 ±0.01
OR OSL 2	2.80	9±2	2.08 ±0.21	5.13 ±0.19	0.95 ±0.03
TW OSL1	3.00	2±1	0.93 ±0.08	1.93 ±0.08	0.20 ±0.01
TW OSL2	5.00	2±1	1.50 ±0.15	5.65 ±0.20	0.98 ±0.03
TW OSL3	4.50	2±1	1.72 ±0.13	4.81 ±0.17	0.86 ±0.03
LEI IV OSL1	3.10	3±1	2.16 ±0.19	6.10 ±0.22	0.91 ±0.03
LEI IV OSL2	2.65	2±1	1.63 ±0.14	4.41 ±0.16	0.65 ±0.02
LEI IV OSL3	2.00	4±1	1.46 ±0.13	6.62 ±0.22	1.05 ±0.03
LEI VI OSL3	2.60	4±1	1.46 ±0.13	5.36 ±0.18	0.84 ±0.03

3. Equivalent dose determination

3.1 Sample preparation

All samples were prepared under subdued red light in the luminescence dating laboratory of the Institute for Geography, University of Cologne. After drying, the fine sand fraction (100-200 μm) of all the samples was separated from the bulk samples by dry-sieving. Samples were then treated with hydrochloric acid in order to remove carbonates. Most of the material dissolved after this treatment and only a few samples yielded sufficient material for further processing. Any clay and organic material was removed by sodium oxalate and hydrogen peroxide. Subsequently, the fine sand quartz fraction was separated from the bulk samples using sodium polytungstate solutions with the specific gravities of 2.68 and 2.62 g/cm^3. In order to purify and further concentrate the quartz fraction and to remove the outer -irradiated layer, the samples were etched in 40% HF for forty minutes, and subsequently rinsed in HCl (10%) to remove fluorides. Even after all the preparations outlined were repeated, only a few samples yielded sufficient quartz grains for OSL measurements. However, in order to prepare the aliquots for the measurements from the remaining samples, about 150 to 300 grains were mounted on stainless steel discs using silicone oil. The number of discs that were measured per sample is listed in Table 2.

The fine-grained fraction was also prepared for dating among all the samples. The grain sizes, from four to eleven micrometres, were enriched by dry-sieving (<40μm) and subsequent settling. Samples were also treated with hydrochloric acid, hydrogen peroxide, and sodium oxalate to remove carbonates, organic material, and clay. Density separation is not suitable for the fine-grained fraction for obtaining a pure quartz and feldspar fraction. Therefore, dating studies were carried out here on the polymineral material. To prepare aliquots for the luminescence measurements, a thin layer of the polymineral fine-grained material was deposited on one centimetre wide aluminium discs after suspension in acetone.

3.2 Equipment and experimental details

All luminescence measurements were carried out on automated Risø TL/OSL readers (TL-DA-12 or 15)[14] equipped with ^{90}Sr/^{90}Y -sources for irradiation. For optical stimulation of the polymineral fine-grained samples, infra-red light emitting diodes (880 ± 80 nm) were used and for stimulation of the quartz samples, blue LEDs (470 ± 30 nm) were employed. For detection of the luminescence emissions, EMI 9235

[14] Bøtter-Jensen *et al.* 2003.

Table 2: Results of dose rate (D_0) calculation, which is based on the values presented in Tab. 1, and equivalent dose (D_e) estimates based on the median of each dose distribution. n is the number of aliquots measured for each sample, and RSD the relative standard deviation of the D_e distribution. All errors represent the 1-sigma uncertainty, with the D_e errors including 5 % uncertainty from the beta source calibration.

Lab.-code	Mineral fraction	Sample	D_0 (Gy/ka)		D_e (Gy)		RSD (%)	n	Age (ka)	
C-L2010	quartz	TC 1	2.51	±0.08	8.82	±0.70	15	21 (22)	3.51	±0.30
	polymin. fine grain		3.29	±0.21	9.17	±0.49	10	11	2.79	±0.23
C-L2011	quartz	TC 2	2.67	±0.09	8.37	±0.59	27	15	3.13	±0.24
	polymin. fine grain		3.50	±0.22	8.78	±0.96	18	10	2.51	±0.32
C-L2012	quartz	TC 3	2.46	±0.08	5.07	±0.34	15	14 (15)	2.06	±0.15
C-L2013	polymin. fine grain	OR1	1.36	±0.11	34.5	±1.8	6	9	25.4	±2.5
C-L2014		OR2	2.41	±0.19	16.02	±0.93	8	8	6.65	±0.65
C-L2008		TW2	2.42	±0.17	12.17	±1.79	24	10	5.03	±0.82
C-L2009		TW3	2.30	±0.17	6.41	±0.42	19	11	2.79	±0.28
C-L2015		LeilV 1	2.68	±0.21	77	±4	5	9	28.7	±2.7
C-L2016		LeilV 2	2.04	±0.16	60	±4	10	9	29.5	±2.9
C-L2017		LeilV 3	2.58	±0.17	6.62	±0.61	15	9	2.57	±0.29
C-L2018		LeiVI 3	2.22	±0.16	8.97	±0.62	8	10	4.04	±0.40

photomultiplier tubes were used with two different detection windows: one for quartz samples in the UV (290 to 370 nm) and the other for the feldspar dominated signals of the polymineral samples in the violet to blue wavelength range (410 ± 30 nm).

The measurement parameters have been determined based on the Single Aliquot Regenerative (SAR) dose protocols described by Murray – Wintle for quartz,[15] and Wallinga *et al.* for feldspars.[16] For all mineral fractions preheat plateau tests and dose recovery tests (DRT; where aliquots were artificially bleached and irradiated with a defined laboratory -dose prior to luminescence measurements) provide the basis for setting the measurement parameters. The results will be presented and discussed in the following section.

4. Measurements and results

4.1 Sand-sized quartz samples

All quartz samples were measured according to the SAR protocol[17] using four different regeneration doses. For optical stimulation, the samples were exposed to blue light for forty seconds and a temperature of 125°C was maintained constantly during illumination. A thermal pre-treatment of 220°C for ten seconds and a test dose cut heat of 160°C were chosen following the methods laid out by Deckers – Vandenberghe.[18] To minimise any effect of recuperation, an optical stimulation at a high temperature (forty seconds at 280°C) was carried out after the measurement of the test dose response curve. This SAR protocol is considered to yield reproducible results for the samples investigated in this study as proved by DRT results. For the DRT, fifteen aliquots of sample TC 1 were artificially bleached and irradiated with a known laboratory dose of 10.9 Gy and subsequently measured using the same SAR procedure as for the natural aliquots. The ratio between the given dose and the measured equivalent dose was 0.92 with the data showing a relative standard deviation (RSD) of 7.5%.

---

[15] 2000; 2003.

[16] 2000.

[17] Murray – Wintle 2000; 2003.

[18] 2007.

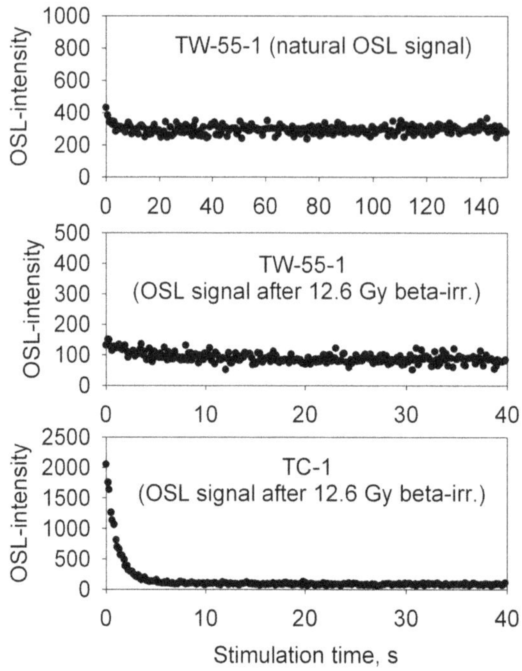

Fig. 1: OSL sensitivity of samples TW 1 and TC 1.

Although sample TW 1 yielded sufficient coarse-grained material for dating, no equivalent doses could be obtained. As illustrated in Figure 1, the sample showed no OSL sensitivity. Even after irradiation with 12.6 Gy the sample showed no real response, in contrast, for example, to sample TC 1. After the same beta dose a typical shine-down curve was obtained for this sample (see fig. 1). TW 2 showed slightly increased sensitivity and sample OR1 even revealed a good dose response. Unfortunately, the quartz coarse-grained material obtained after sample preparation was insufficient for measuring a substantial number of aliquots.

Finally, OSL ages were only obtainable for the coarse-grained quartz fraction among the samples from the Tell Brak profile (OSL samples referred to as TC). The results of the OSL measurements are summarised in Table 2 and the equivalent dose distributions are shown in Figure 2. With respect to the data spread, the RSD of 15% (note that the outlier as shown in the histogram was excluded from the dataset for equivalent dose calculation) for samples TC 1 and 3 is considered to be in the range of well-bleached samples given the fact that the fully bleached and homogeneously irradiated dataset of the dose recovery test already showed a RSD of 7.5%. Sample TC 2 shows a larger dispersion of the individual values (RSD 27%), but the values are still normally distributed. The equivalent dose values calculated from these distributions are based on the median. This value is a good representative of the peak of the distributions for all the samples. The errors within the median and common age equivalent dose values are consistent. Nonetheless, according to the comparatively small number of aliquots measured per sample, the median was considered to be the best mean value.

4.2 Polymineral fine-grained samples

As the luminescence emission of feldspars is much brighter than that of quartz in polymineral samples, the emission is dominated by the feldspar signals. Hence, the polymineral fine-grained samples were measured using the protocol based on the proposition of Wallinga *et al.*[19] for sand-sized K-rich feldspars. Optical stimulation was carried out for 350 seconds using IR-light emitting diodes with the samples held at 30°C throughout illumination. Figure 3 illustrates the pre-heat and dose recovery pre-heat plateau tests carried out for sample TC 1.

A thermal pre-treatment of 220°C for ten seconds was chosen based on the results of the pre-heat plateau tests shown in Figure 3. By using a pre-heat temperature of 220°C the natural samples (fig. 3a) show the best reproducibility and comparatively low recuperation values. When bleached and irradiated with a defined laboratory dose (fig. 3b), the discs that were pre-heated to 220°C showed excellent recovery of the given dose. Furthermore, a good reproducibility among the individual discs and low recuperation

[19] 2000.

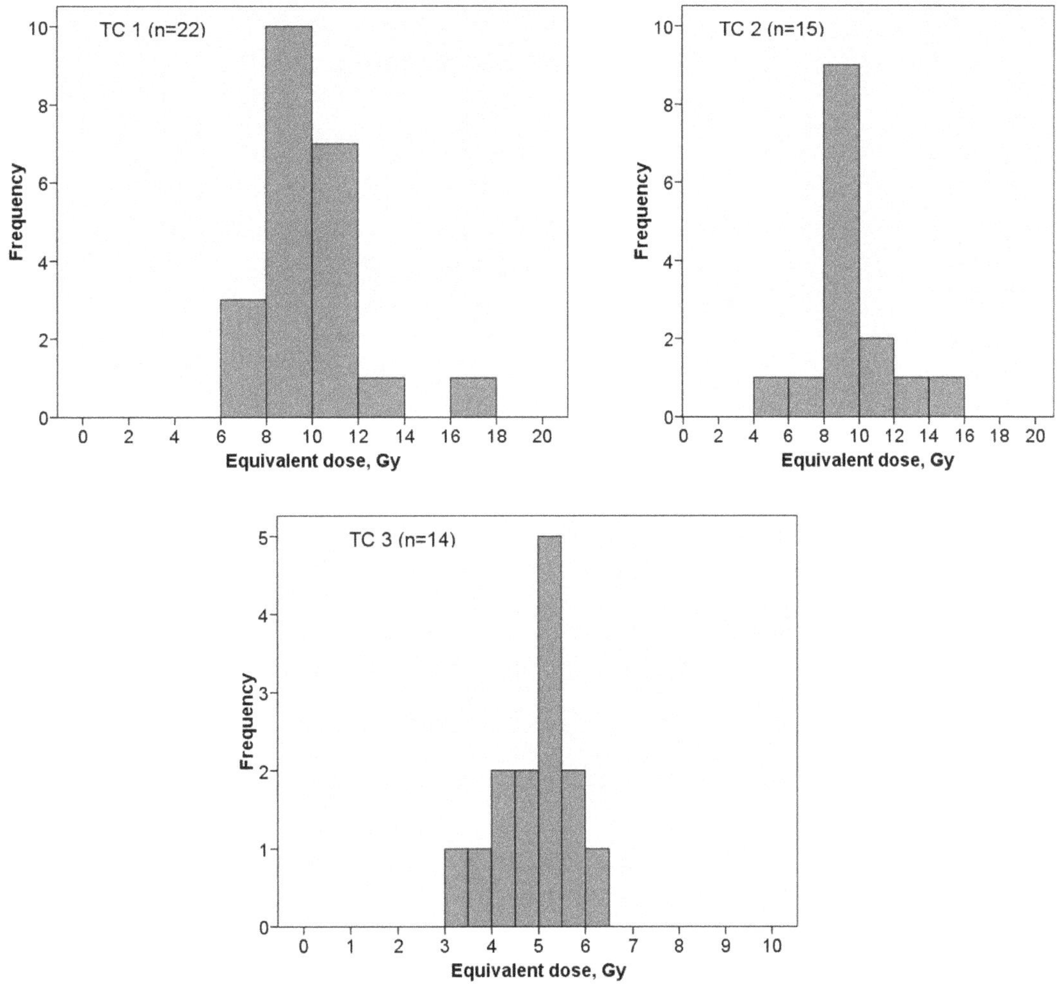

Fig. 2: Equivalent dose distributions for samples TC 1 (a), 2 (b), and 3(c).

effects were observed. A dose recovery test using nine aliquots of sample TC 1 was also carried out with pre-heating at 220°C for ten seconds. As the ratio of administered to measured dose was 0.98, the SAR protocol was considered to yield reproducible results for the samples investigated in this study. The results of the IRSL measurements are summarised in Table 2.

Samples TW 2 and 3 showed dim signals. This might explain the high data spread observed in the equivalent dose distributions of these two samples. With twenty-four and nineteen percent, these samples yielded the highest RSD values. The fully bleached and homogeneously irradiated aliquots of sample TC 1 used for the DRT showed a RSD of 7.4%. Some of the natural dose distributions yielded even smaller values indicating that the luminescence characteristics of the individual samples have a strong impact on this parameter. The fluvial samples might be overestimated due to insufficient bleaching by sunlight during transport. Fine-grained samples are not as good as coarse-grained samples to estimate the degree of resetting as several thousands of grains always contribute to the luminescence signal, which will then be dominated by the brightest ones. In most cases of poorly bleached sediments, the brightest grains are often those carrying the highest residue. The fact that samples from fluvial deposits (sections OR and LEI IV) largely show quite low RSD values might be interpreted as evidence for sufficient bleaching prior to deposition. As preparation of the coarse-grained fraction did not yield sufficient material for dating, additional DRTs could perhaps further clarify this question.

The major problem encountered in feldspar dating is a signal loss over time, a phenomenon called *anomalous fading*,[20] which affects feldspars from at least some geological origins and causes an

[20] Wintle 1973.

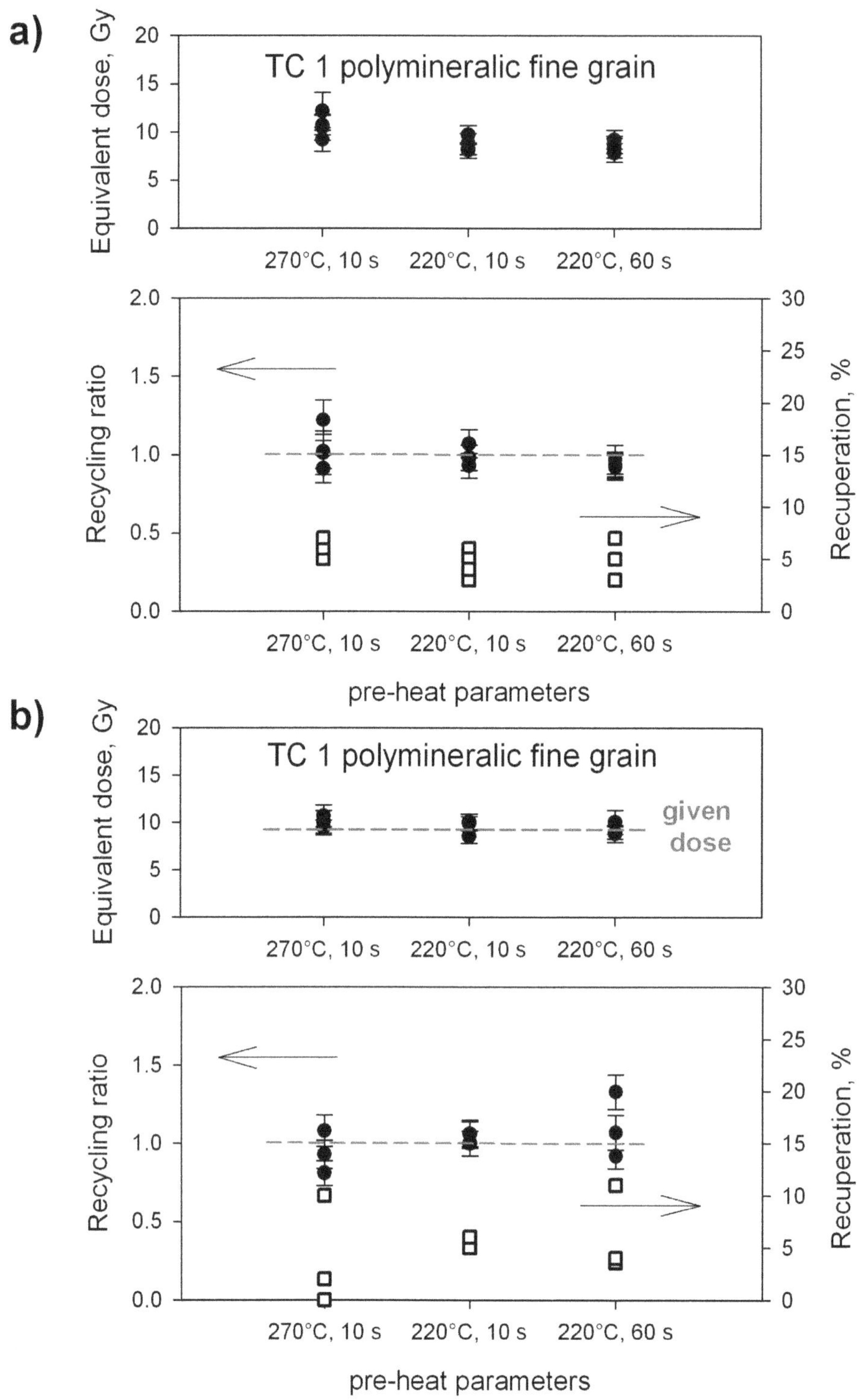

Fig. 3: Dependence of the equivalent dose, the recycling ratio and the recuperation effect on varying preheat temperatures. a. Pre-heat plateau test for sample TC 1 (fine grain). Nine natural aliquots were measured in groups of three using different thermal pre-treatments. The open squares represent the corrected OSL signal due to recuperation observed after administering the zero dose in the fourth regeneration cycle (expressed in % of the corrected natural signal). This recuperation value should not exceed a threshold of c. 5 % to prevent erroneous D_e estimates. b. 'Dose recovery' pre-heat plateau test measured for sample TC 1 (fine grain). All measurement conditions were the same as described for graph a, but all nine aliquots were bleached prior to beta irradiation with the 'recovery dose' (indicated by the grey dashed line for clarity). Subsequently, the aliquots were measured using the SAR protocol with different pre-heating treatments.

underestimation of the luminescence age if this is left uncorrected. Despite a few studies reporting on volcanic quartz fading, it is not known to affect quartz samples.[21] Testing feldspar samples for the occurrence of fading is necessary. But in general the results of such experiments are more qualitative than quantitative. Therefore, it is difficult to correct for fading unless independent age control is available. Nevertheless, several procedures have been suggested that provide a test for the occurrence of fading and correct for the signal loss, if present.[22]

Samples TC1 and 2 already provide evidence for the presence of a fading problem. For both samples the ages of the fine-grained fraction are significantly younger than the quartz ages (see tab. 2). The ratio of quartz versus fine-grained material for TC 1 and 2 is 1.26 and 1.25, respectively. Thus, a similar rate of underestimation might affect all other ages obtained for fine grains as well; although, this should be tested in detail. The interpretation of all ages of fine-grained material should be considered as minimum ages and about 25% might be added to these ages.

Bibliography

AITKEN, M.J.
1985 Thermoluminescence dating.
1998 An introduction to optical dating; The dating of quaternary sediments by the use of photon-stimulated luminescence.

AUCLAIR, M. – LAMOTHE, M. – HUOT, S.
2003 Measurement of anomalous fading for feldspar IRSL using SAR, Radiation Measurements 37, 487-492.

BØTTER-JENSEN, L. – MCKEEVER, S.W.S. – WINTLE, A.G.
2003 Optically stimulated luminescence dosimetry.

DECKERS, K. – VANDENBERGHE, D.
2007 Confronting luminescence with radiocarbon dates for fluvial deposits in the Upper Khabur Basin of northeastern Syria. In: Ooghe, B. - Verhoeven, G. (eds.), Broadening Horizons - Multidisciplinary Approaches to Landscape Study, 50-74.

DULLER, G.A.T.
2004 Luminescence dating of Quaternary sediments: recent advances, Journal of Quaternary Science 19, 183-192.

HUNTLEY, D.J. – LAMOTHE, M.
2001 Ubiquity of anomalous fading in K-feldspars and the measurement and correction for it in optical dating, Canadian Journal of Earth Sciences 38, 1093-1106.

LAMOTHE, M. – AUCLAIR, M. – HAMZAOUI, C. – HUOT, S.
2003 Towards a prediction of long-term anomalous fading of feldspar IRSL, Radiation Measurements 37, 493-498.

LIAN, O.B. – ROBERTS, R.G.
2006 Dating the Quaternary: progress in luminescence dating of sediments, Quaternary Science Reviews 25, 2449-2468.

MURRAY, A.S. – WINTLE, A.G.
2000 Luminescence dating of quartz using an improved single-aliquot regenerative-dose protocol, Radiation Measurements 32, 57-73.
2003 The single aliquot regenerative dose protocol: potential for improvements in reliability, Radiation Measurements 37, 377-381.

PRESCOTT, J.R. – HUTTON, J.T.
1988 Cosmic ray and gamma ray dosimetry for TL and ESR, Nuclear Tracks and Radiation Measurements 14, 223-227.
1994 Cosmic ray contributions to dose rates for Luminescence and ESR Dating: large depths and long-term variations, Radiation Measurements 23 (2/3), 497-500.

[21] e.g. Duller 2004.

[22] see e.g. Huntley – Lamothe 2001; Auclair *et al.* 2003; Lamothe *et al.* 2003.

PREUSSER, F. – ANDERSEN, B.G. – DENTON, G.H. – SCHLÜCHTER, C.
2005 Luminescence chronology of Late Pleistocene glacial deposits in North Westland, New Zealand, Quaternary Science Reviews 24, 2207-2227.

PREUSSER, F. – DEGERING, D. – FUCHS, M. – HILGERS, A. – KADEREIT, A. – KLASEN, N. – KRBETSCHEK, M. – RICHTER, D. – SPENCER, J.
2008 Luminescence dating: Basics, methods and applications, Eiszeitalter und Gegenwart/ Quaternary Science Journal 57/1-2, 95-149.

WALLINGA, J. – MURRAY, A.S. – WINTLE, A.G.
2000. The single-aliquot regenerative-dose (SAR) protocol applied to coarse-grain feldspar, Radiation Measurements 32, 529-533.

WINTLE, A.G.
1973 Anomalous fading of thermoluminescence in mineral samples, Nature 245, 143-144.

12. The "dung-as-fuel" model tested at two Syrian Jezirah sites

Katleen Deckers

1. Introduction

It is often suggested that dung was used as fuel in the Near East. Such interpretations are frequently based on the charred seed-to-charcoal proportions.[1] This model assumes that if most of the seeds come from dung fuel, then the wild seed-to-cultivated-cereal ratio is an indicator of what the herds and flocks ate.[2] Sometimes, as a consequence of this model, all burned botanical remains from archaeological sites within this region are considered to have derived from burned dung. Furthermore, it is argued that the animals from this same area were foddered with cereals. This is supported by rare texts that mention such practises but should not be seen as the rule because the model has several flaws.[3] The main problem with the model is that it cannot discern between charred seeds derived from dung and those from crop-processing or food. Another problem with the model is that experimental work has demonstrated that cereal grains with large seeds (especially the free-threshing cereals) do not preserve very well when compared to small seeds within goat/sheep coprolites.[4]

The charred seed-to-charcoal proportion is often used as a proxy for the availability of fuel in the Near East. It is true that the use of dung as fuel is often linked with deforestation.[5] However, ethnobotanical research also has shown that sometimes dung is preferred as fuel and not necessarily indicates a lack of fuel.[6] For example, in the Jebala of Morocco, where woodland is still abundant, dung is used as fuel for making pottery because the type of dung used achieves a higher concentration of heat and avoids quick cooling; its combustion is slower and more regular than with wood.[7] It should be noted, however, that fuel properties of dung change according to the type of dung used and the way it was prepared for fuel. Although pieces of animal dung are sometimes burnt directly, dung is usually burnt in the form of cakes produced by mixing it with other plant material, especially with crop processing by-products.[8] Another reason dung may have preferentially been used as fuel may be due to its relative ease of preparation.[9]

Up until recently, little was known about the vegetation development in northern Syria because of the poor preservation of pollen and the low chronological resolution of existing pollen cores.[10] Nonetheless, as revealed through anthracological research at several sites in southern Turkey and northern Syria, much information has been gained concerning the development of Bronze Age vegetation.[11] These data provide a regional diachronic insight that can be used to assess the necessity of using dung as fuel. Additionally, further information used in evaluating the density of vegetation exists for the Upper Khabur region, where the intensity of agriculture for the third millennium BC was investigated. Findings from this research reveal some information on how much woodland could have been preserved in this area.[12]

Within the study reported on here, the seed-to-charcoal proportions from the sites of Tell Mozan in the Upper Khabur Basin, and Emar in the Jezira, are investigated and compared against their qualitative charcoal data. Furthermore, for the Tell Mozan region, data are investigated that indicate the amount of agricultural land during the third millennium BC (fig. 1).

---

1 Miller 1984; 1986; Miller – Smart 1984; Charles 1998.

2 Miller 1984; 1996.

3 Pruß – Sallaberger 2005.

4 see e.g. Anderson – Ertug-Yaras 1998; Valamoti – Charles 2005.

5 e.g. Bottema 1984; Miller 1984; Anderson – Ertug-Yaras 1998.

6 e.g. Barnard – Kristoferson 1985; Penã *et al.* 2003.

7 Penã *et al.* 2003.

8 Barnard – Kristoferson 1985.

9 Barnard – Kristoferson 1985.

10 Bottema 1989.

11 Deckers – Pessin, chapter 3 this volume.

12 Deckers – Riehl 2008.

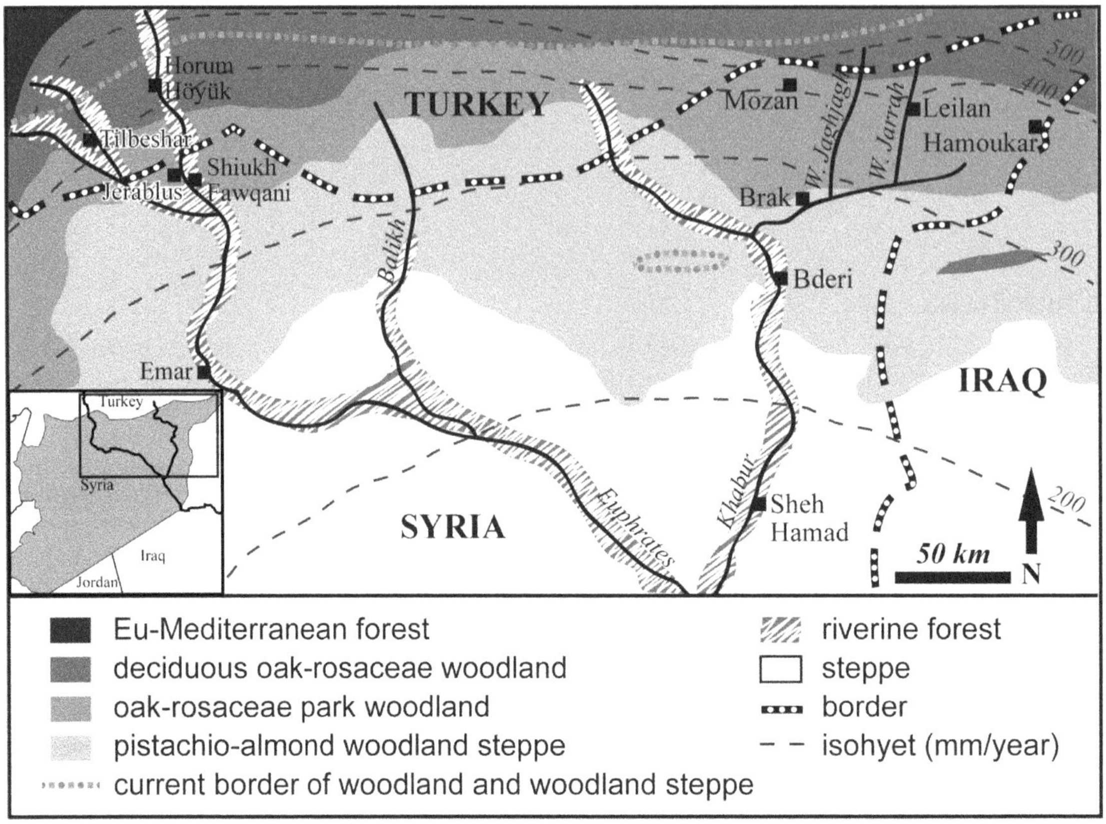

Fig. 1: Potential vegetation map with location of sites (after Hillman in Moore et al. 2000).

2. Investigated sites: location, history and palaeoclimate

Mozan is located in the Upper Khabur Plains of northern Syria (fig. 1) and lies within a mean annual rainfall zone of around 420 mm/year. It is not situated along a perennial stream and today, hardly any trees grow in the area. Instead, the land is intensively used agriculturally for cereal cultivation. Studies of the potential vegetation, however, indicate that under present day climatic conditions and in the absence of human impact, Mozan would naturally be located in the oak park woodland zone.[13]

Emar (fig. 1) garners a much lower mean annual precipitation, at between roughly 200 and 300 mm, but in dry years (e.g. between 1958 and 1960) this zone lies near the 100 mm limit. Although it welcomes less rainfall, the site is located along the Euphrates, which must have played an important role. Considering the low level of annual rainfall, irrigation may have been necessary for cultivation of fields in the neighbourhood. The potential vegetation of the Middle Euphrates would have been lusher than it appears today, especially the riverine vegetation; but also away from the Euphrates, pistachio-almond woodland steppe would have occurred if no human interference had taken place.[14]

The Mozan samples represent the Early Bronze Age (EBA) and early Middle Bronze Age (MBA), while the Emar samples cover a longer stage, from the EBA to the Late Bronze Age (LBA). During the period under consideration, a general trend of climatic drying was observed, although fluctuations did occur. In comparison with the present climate, the MBA was probably just as dry as today.[15]

3. Methods

In order to gain insight into the evolution of vegetation in the investigated sites' surroundings, charcoal fragments have been analysed and results will be detailed here mainly based on fragment percentages and ubiquity percentages. Special attention was paid to record the presence of small twigs, which are indicators

[13] Moore *et al.* 2000.

[14] Hillman in Moore *et al.* 2000.

[15] e.g. Lemcke – Sturm 1997; Wick *et al.* 2003.

of the local taxa present. In total, 27,584 fragments from forty-five Emar samples and 9,972 fragments from 175 Mozan samples have been identified to reconstruct the vegetation.

Most of the identified charcoal samples derive from floated samples at Emar, while at Mozan there is an additional proportion derived from hand-picked sediment samples that were not floated. For calculating the seed-to-charcoal ratios only floated samples were used and posthole contexts were excluded from the analysis. An exception is sample 60/51-89 from Emar, where there exist some reservations as to whether it really represents a posthole context given the variety of botanical remains found in the associated sample. Although seed-to-charcoal ratios are often based on weighed fragments,[16] this investigation is based on counted fragments. The seed and/or fruit count data derive from Dr. Simone Riehl and has been published elsewhere in detail.[17]

In order to gain additional information about the density of the former vegetation, resource calculation data are used. The methodology used in procuring these data is published in detail in Deckers and Riehl,[18] and in Deckers and Drechsler.[19] In summary, tell and lower town sites assumed to have had their maximal occupation in the third millennium BC were mapped from CORONA satellite images in a GIS system. Based on the mapped sections, the sustaining areas for this region during the Early Bronze Age have been calculated using realistic population density values, food consumption patterns, and agricultural yields within this climatic zone. A standard population density within this region normally is about 200 persons/ha.[20] An extrapolation of food consumption patterns from 1970s Syria to the archaeobotanical data made it possible to calculate land needed to feed the people living in this area. This was done using yield data from the Food and Agriculture Organization of the United Nations (FAO) for the 1961-1970 period and taking into account settlement size. For the required amount of seed for the next harvest, an extra 6% was added.[21]

4. Results

4.1 Tell Mozan

4.1.1 Vegetation reconstruction based on charcoal from Tell Mozan

Deciduous oak (*Quercus* spp.) dominates the Mozan charcoal assemblage, representing between 27% and 82% of the identified fragments for all periods (fig. 2). As a whole, *Quercus* is present in 60.5% to 94.6% of the samples (fig. 3). The presence of acorns and small oak twigs amongst the botanical remains confirms the presence of deciduous oak woodland in the region. The most striking trend within the charcoal data from Mozan is the decline in fragment percentages and oak ubiquity in the MBA strata as compared with earlier levels. While oak is present in 92% of the Early Jezirah (FG[22]) III samples (2700-2250 BC) and 95% of the FGIV samples (2250-2100 BC), it only appears in 70% of the FGV (2100-2000 BC) samples and in 61% of the Old Jezirah (AG[23]) I-II samples (2000-1800 BC). Of the identified charcoal fragments, 70% of those from FGIII, 60% from FGIV and 64% from FGV strata consist of deciduous oak. However, in the MBA (AGI-II) samples, only 27% of the identified fragments are from deciduous oak.

The Maloideae (a Rosaceae subfamily), which typically would have accompanied this open oak park woodland, are also less frequently found in samples from FGV and AGI-II strata. However, a contradicting trend is observable in fragment percentage charts, which indicates that there were relatively more Maloideae fragments identified from FGV and AGI-II strata.

Within the MBA strata, olive appears for the first time at the site, being present in 12% of the samples and consisting of 39% of the fragments. Another pattern within the data is the decrease of poplar (*Populus* spp.) and/or willow (*Salix* spp.) in FGV samples in relative numbers, as well as in relative presence. They are extant in 47% of the FGIII strata samples, in 40.5% of the FGIV samples, but in only 9% of the FGV samples and again in 21% of the AGI-II samples. A similar pattern can be observed for plane (*Platanus* spp.), which was present in 12% of the FGIII samples, 16% of the FGIV samples, and in only 9% of FGV and 5% of MBA samples. However, the charcoal fragment percentages for *Platanus* do not show an entirely consistent trend. While they are present in less MBA samples, the charcoal fragment percentages are

[16] see e.g. Miller 1984; 1996; Miller – Smart 1984.

[17] Riehl 2010a; Riehl 2010b.

[18] Deckers – Riehl 2008.

[19] Deckers – Drechsler, chapter 2 this volume.

[20] see Kramer 1980.

[21] detailed in Deckers – Riehl 2008.

[22] The abbreviation derives from the original terminology "Früh-Jezirah."

[23] The abbreviation derives from the original terminology "Alt-Jerzirah."

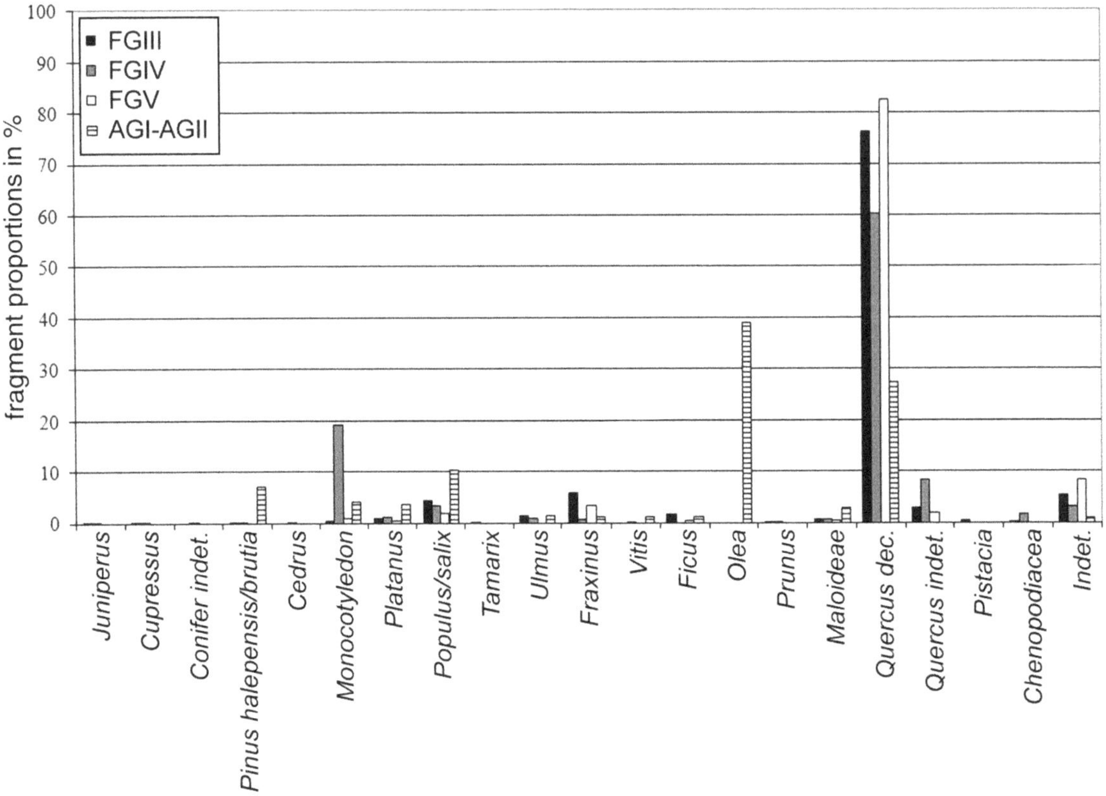

Fig. 2: Charcoal identification results from Tell Mozan in fragment percentages.

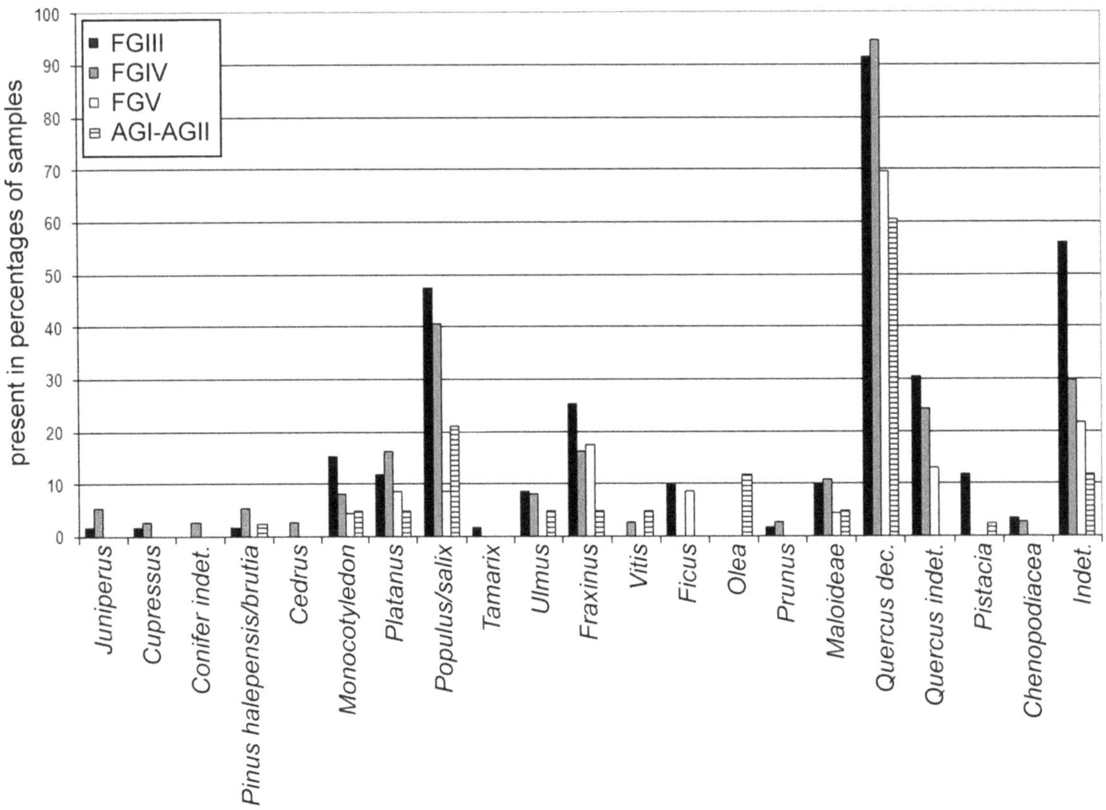

Fig. 3: Charcoal identification results from Tell Mozan as ubiquity percentages from Mozan.

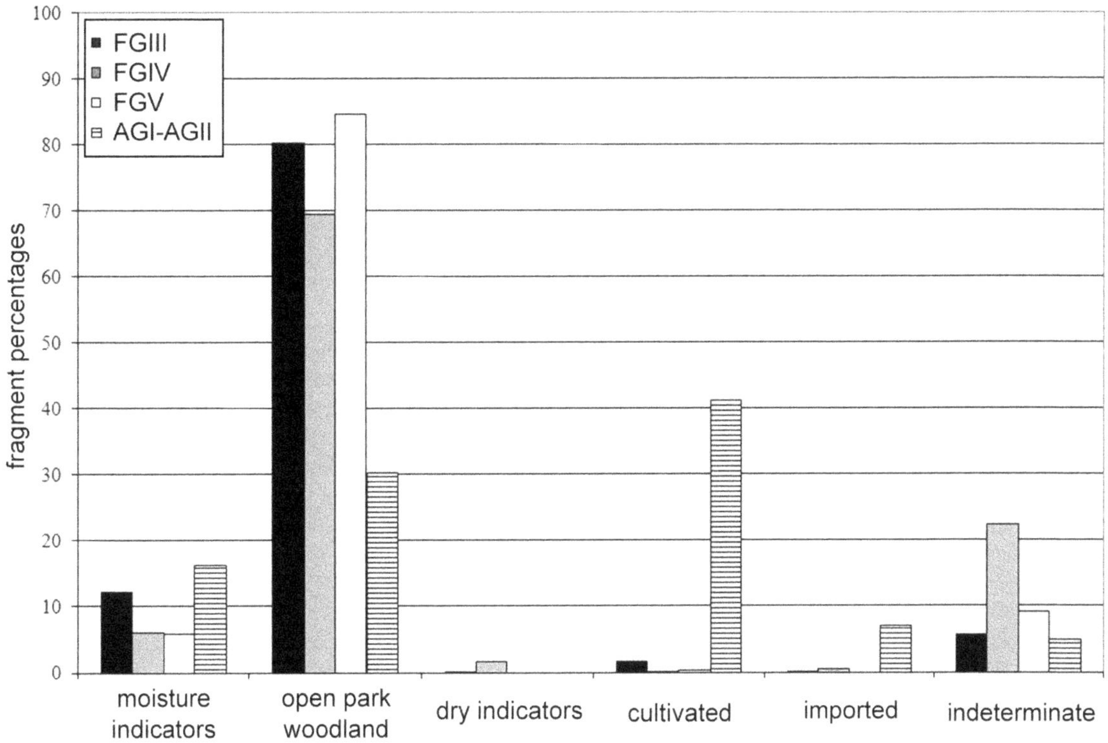

Fig. 4: Summary of charcoal results from Tell Mozan; fragment percentages in groups.

higher in the MBA strata. *Platanus* is present within the samples in small numbers and only consists of up to 1.3% of the identified fragments. Elm (*Ulmus* spp.) shows a ubiquity trend similar to *Platanus*. In the FGIII and FGIV strata, it is present in 8% of the samples, while it is entirely not found in FGV samples and then appears again in 5% of the AGI-II samples. Chenopodiaceae, are present in relatively few FGIII and FGIV samples, and non-existent in FGV and AGI-II samples.

It is striking that in the MBA olive (*Olea* spp.) appears at Tell Mozan in relatively large numbers whereas *Quercus* decreases within the samples. While olive trees grow naturally in the Mediterranean area,[24] the *Olea* charcoal finds from MBA strata at Mozan are clearly found outside of their natural distribution range. This suggests that *Olea* was either cultivated or then imported. At present, no olive stones have been found in the archaeological strata at Tell Mozan. The earliest *Olea* remains located outside their natural distribution area and near Mozan were found in the Late Chalcolithic levels at Hassek Höyuk along the Euphrates. This suggests that *Olea* might have been cultivated at this time.[25] However, cultivated olive wood and stones cannot be distinguished from wild varieties. *Olea* wood remains were also found in the EBA I levels of Horum Höyük and were recorded in larger proportions in the EBA III strata from Tilbesar.[26] The expansion of *Olea* in the MBA levels from Mozan may reflect an intensive cultivation of the species at this time. Alternatively, the charts may indicate that there were proportionally fewer oaks growing in the surroundings of Mozan, which could have been a result of climatic changes. One should note that Tell Mozan is located on the northern fringes of the presumed zone of potential oak-Rosaceae park-woodland. In order for the vegetation zone to have changed this dramatically, a drastic drying event would have been necessary. However, it seems more likely that a combination of both climatic drying *and* human impact may have been responsible for the recorded relative decline of oak fragments in the archaeological samples.

Because some of the taxa are present in relatively small numbers, it is questionable whether or not the trends observed are representative. Therefore, the taxa have been grouped according to habitats (fig. 4). Oak, pistachio, Rosaceae species, and juniper were considered as belonging to open park woodland, while plane, poplar, willow, elm, ash, and tamarisk were interpreted as originating from riverine habitats. Moreover, Chenopodiaceae are considered to be dry indicators, while conifers, excluding juniper, were almost certainly imported. The cultivated species probably consisted of vine, olive trees, and fig

[24] Zohary – Hopf 2000.
[25] Gregor 1992.
[26] Pessin 2004.

trees. The monocotyledons and unidentifiable fragments are summarised as "unspecifiable." While open park woodland decreased in the MBA, there is a clear signal for an increased number of fragments originating from cultivated taxa while more moisture indicators and imported taxa are also accounted for. Alternatively, the FGIV and FGV have the least amount of moisture indicators. As a whole, most drought-indicators were found in the FGIV strata. The diminution of moisture-indicating taxa in FGIV and FGV samples might be related to drier climatic conditions, while the peak in moisture-indicating taxa at about 2000-1800 BC could indicate moister climatic conditions. However, the largest changes in woody vegetation proportions and ubiquity are visible between the EBA and early MBA, when human impact appears to have played a role.

4.1.2 Charcoal-to-seed proportions from Tell Mozan

Looking at the charcoal-to-seed proportions from Tell Mozan it is noted that charcoal only constitutes a minor proportion in all periods, although there are some fluctuations visible over time (e.g. hardly present in the FGIII flotation samples, at hand in larger percentages in the FGIV and FGV periods, equally low charcoal percentages in the AGI-II and the FGIII samples) (fig. 5). The figure of wild-to-cultivated seed proportions from the same contexts[27] further indicates that wild seeds are, for the most part, dominant in the samples (fig. 6). While these ratios suggest the use of dung as fuel, only two goat and/or sheep coprolites were found within the samples. Interestingly, one of these was found in association with an oven (MZ01 BP36 - q1118 - Fs1519; fig. 7) that contained more than 85% of seed and fruit fragments. The other two oven context samples (MZ01 BP53-q2156-Fs1925 and MZ01 BP68-q1199-Fs1541) show some variety in terms of the general composition of their botanical remains. Whereas oven MZ01 BP53 (q2156-Fs1925) contains a considerable amount of charcoal, oven MZ01 BP68 (q1199-Fs1541) does not. Although no coprolites were found within the samples, the high number of wild seeds within both suggests dung was used as fuel.

The most numerous of the wild plant seeds were those from *Poa* spp.[28] These may have entered the botanical record in connection with animal dung, although an alternative explanation would be to consider their presence as crop-processing by-products.[29] Most of the remaining wild plant seeds arrived at the site as crop weeds, either directly or indirectly via animal dung.[30] Riehl[31] only found a few taxa in the Tell Mozan remains that are traditionally known to be cut and stored as winter fodder in Anatolia.

How do we explain the presence of deciduous oak park woodland in the neighbourhood of Tell Mozan when the possible use of dung as fuel would suggest a lack of woody resources in the area? Either dung was the preferred fuel or the deciduous oak park woodland vegetation was not very dense due to intensive cultivation in the Upper Khabur Plains. In order to test the second of these hypotheses, the size of the agriculturally-exploited area within the Upper Khabur Plains during the third millennium BC has been estimated.

4.1.3 Cultivated areas in the Tell Mozan region, third millennium BC

Assuming a population density of 200 persons/ha within the settlements, and depending on the archaeobotanical data used for the calculations, the area under permanent cultivation was estimated to cover between 21.2% and 22.1% of the region under investigation. If one assumes rather population densities of 400 persons/ha were the case, between 42.5% and 44.3% of the area could have been under cultivation.[32] Therefore, even with population densities of between 200 and 400 persons/ha there was still space available for pasture and oak park woodland vegetation. This contradicts what has often been suggested hitherto, namely that the Upper Khabur Basin (particularly towards the south) would have been so densely populated during the EBA that it could hardly have provided enough agricultural land for its inhabitants, nor have allowed for woodland development.[33]

In order to investigate the impact of tribute on the environment, the number of consumers in the population estimates was increased. There exist, for example, indications for the demand of tribute by the Akkadians in the Upper Khabur at around 2300 BC. The best evidence for this southern intervention in Syria was found at Tell Brak where an Akkadian administrative building was found that possibly

---

[27] data from Riehl 2010a.

[28] Riehl 2010a.

[29] Riehl 2010a.

[30] Riehl 2010a.

[31] Riehl 2010a.

[32] Deckers – Riehl 2008.

[33] Wilkinson 2003.

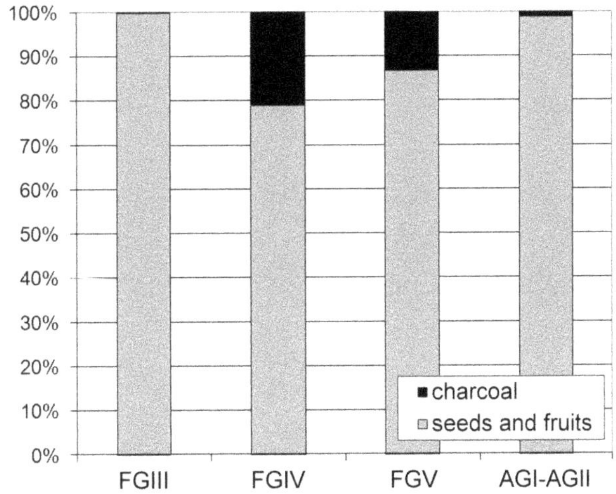

Fig. 5: Charcoal-to-seed proportions from Tell Mozan (based on six samples for the FGIII, four for the FGIV, ten for the FGV, and thirty-one for the AGI-II).

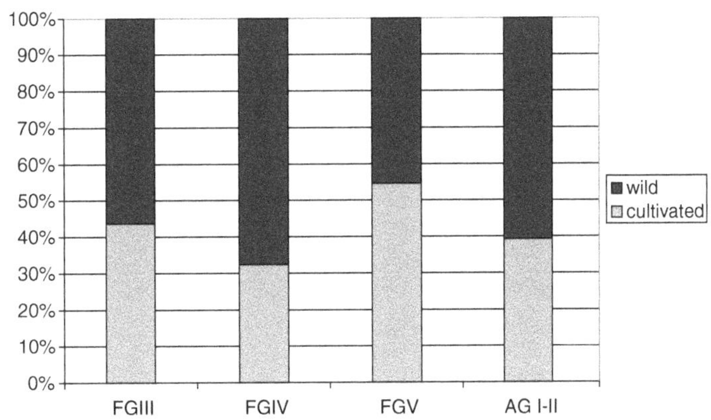

Fig. 6: Wild-to-cultivated seed proportions from Tell Mozan (based on six samples for the FGIII, four for the FGIV, ten for the FGV, and thirty-one for the AGI-II) (data from Riehl 2010a).

was related to the collection of tribute from the region and provisioning of troops. Tell Mozan almost certainly was a powerful Akkadian client- or ally-state.[34]

If this were in fact the case, that tribute had to be provided, the calculations indicate that the resource limits must have been reached in many areas, leaving only between 33.6% and 36.3% of the area uncultivated. Perhaps long-term consequences for this period are visible in the AGI-II samples from Mozan, from which less oak fragments were identified.

4.2 Emar

4.2.1 Vegetation reconstruction based on charcoal from Emar

Populus/Salix clearly predominates at the site of Emar in number and ubiquity. *Tamarix* (tamarisk) is also present in large percentages and in a large number of samples (fig. 8). All other taxa exist in minor proportions. The most striking trend is the decreased proportion of *Populus/Salix* in relation to *Tamarix* in the MBA and its subsequent increase in the LBA again. The high percentages of *Populus/Salix* and *Tamarix*, similar to those of other Middle Euphrates sites, indicate the presence of an extensive riverine forest. Alder (*Alnus* spp.), *Ulmus*, reed (*Phragmites* spp.), and *Platanus*, which were found in Emar in minor quantities, were also probably part of this gallery forest. Today, the Syrian Euphrates vegetation no longer includes

[34] Akkermans – Schwartz 2004.

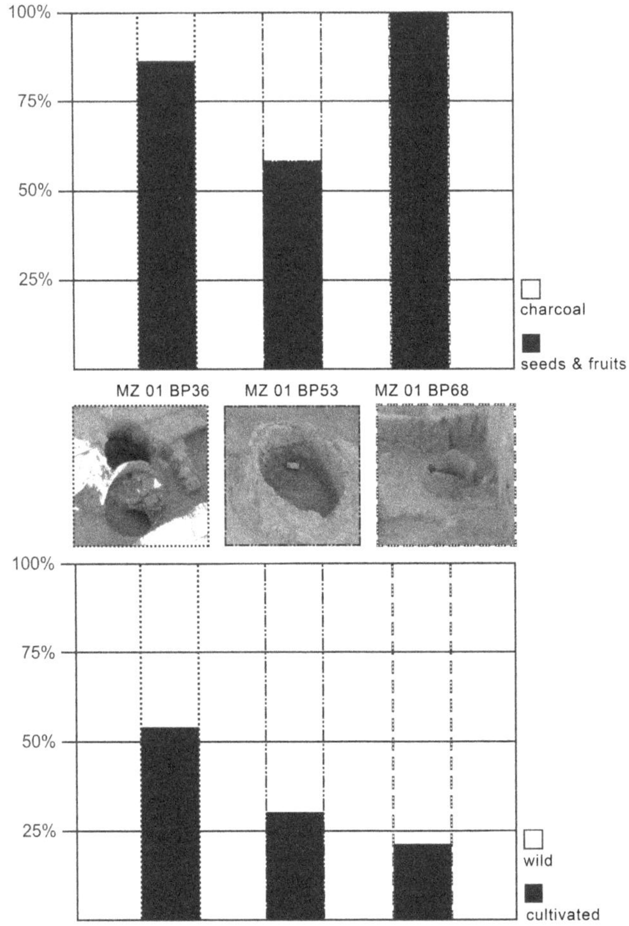

Fig. 7: Charcoal-to-seed ratio and wild-to-cultivated seed ratio for Tell Mozan oven contexts.

some of these species. Hence, during the Bronze Age it was probably more varied and lush in comparison to today. Small twigs of *Populus/Salix*, *Tamarix*, and *Ulmus* have been found which underlines their local presence.

In all Bronze Age phases at Emar, woodland steppe proportions, represented by *Pistacia* non-*atlantica*, *Lycium* (wolfberry), *Ziziphus* (Christ's thorn), and Chenopodiaceae (goosefoot species), are somewhat lower compared to other riverine sites in northern Syria and southern Turkey. More specifically, during these phases the woodland steppe is expressed in values of 0.18 to 2.7% at Emar, while there is circa 10% of woodland taxa at Tell Shioukh Fawqani,[35] circa 4% at Jerablus, and 5.5% at Bderi.[36] The low values for woodland steppe at Emar, however, could indicate that the main zone of land use here was within the floodplain and that this was not the case at the other sites. Wilkinson[37] observes that Emar's location with its catchment mainly within the floodplain is in marked contrast to most Bronze Age settlements farther to the north, which are located on the main Euphrates terrace. The presence of gazelles and onagers at LBA Emar further suggests relatively lush woodland steppe vegetation compared to the present-day vegetation,[38] and supporting the last argument that prime agricultural activities and wood resources were located within the floodplain.

4.2.2 Charcoal-to-seed ratios in the botanical samples from Emar

The average proportion of charcoal-to-seeds within the Emar botanical samples shows a different pattern to the one observed at Mozan. In the latter case, seeds and fruits dominate and charcoal proportions

---

[35] Pessin 2004.

[36] Deckers – Riehl 2007; Engel 1993.

[37] Wilkinson 2004.

[38] Gündem – Uerpmann 2003.

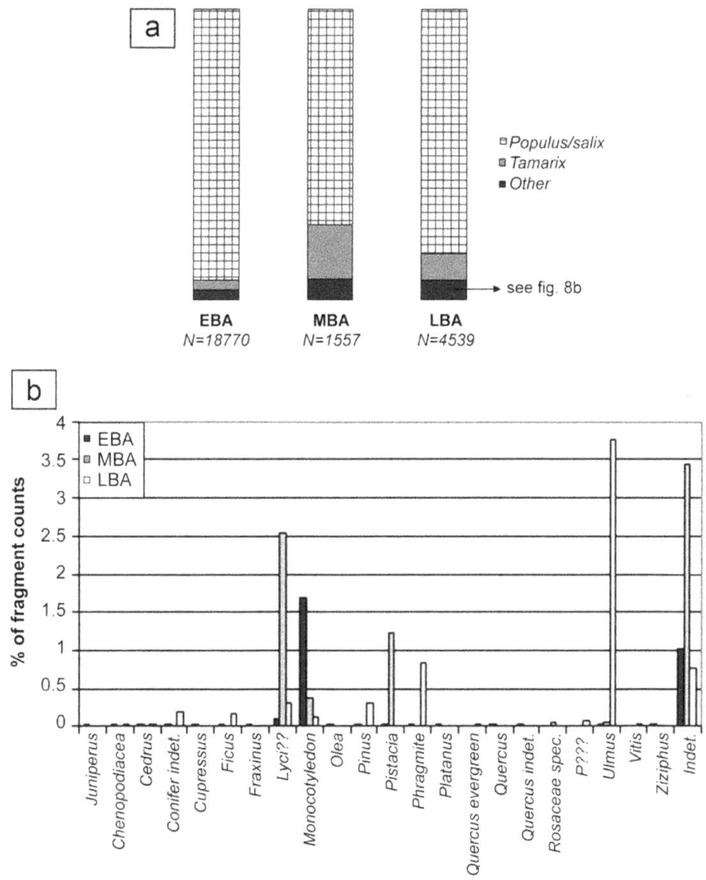

Fig. 8: Charcoal identification results from Emar in fragment percentages. Percentages are based on 18,770 fragments for the EBA, 1,541 for MBA and 4,539 for LBA. 8a. Summary of results; 8b. *Populus/Salix* and *Tamarix* not included in the results.

are rather low but at Emar charcoal proportions clearly outnumber seeds. This indicates that wood was an important source for fuel at Emar (fig. 9).

Nevertheless, there is a diachronic trend within the Emar samples, with a somewhat smaller charcoal proportion in the MBA, as compared to the EBA and LBA. This may indicate an increased use of dung as fuel during the MBA. Moreover, in two of the three samples included in the analysis from the MBA, coprolites are present; one of which even derives from a hearth context. The relatively high wild seed-to-cultivated seed proportions (50:50) also suggests that although wood was used as fuel, dung may have played an additional role in the same function.

When looking in greater detail at the investigated oven and hearth contexts, some variability within the botanical assemblages is noted (fig. 10). Some of these contexts were dominated by charcoal (60/51-95, 75/52-62, and 65/53-53) and clearly show the use of wood as fuel, while others were dominated by wild fruit and seeds (63/50-91 and 76/55-81), which is indicative of the use of dung as fuel and is supported by the find of one sheep/goat coprolite. Nonetheless, it is interesting to note that within numerous oven contexts poplar or willow was the chosen fuel, neither of which are normally considered adequate firewood. Poplar, for instance, burns slowly producing a black choking smoke even when seasoned.

5. Discussion

The data from Mozan indicate that dung played a more important role there as fuel than it did at Emar. However, both sites seem to have used dung as fuel, notwithstanding both sites had access to woody resources in the neighbourhood. The woody resources from Emar were probably somewhat denser than those within the Mozan region, the latter of which can be explained by the type of woody vegetation that occurred. While woodland within the riverine forest must have been considerably dense, oak park woodland contained less woody resources per square metre.

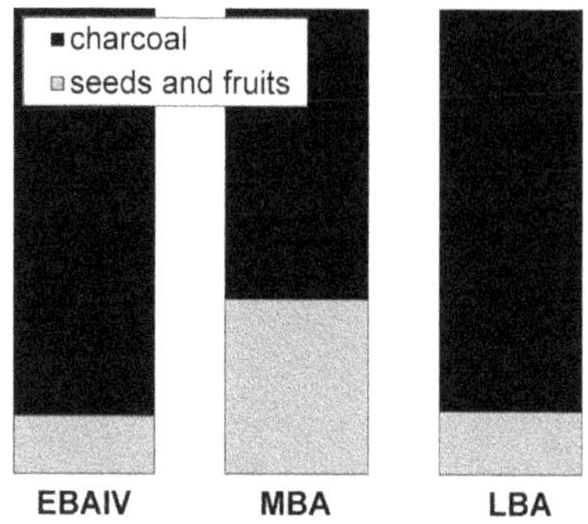

Fig. 9: Charcoal-to-seed/fruit proportions for EBA, MBA, and LBA samples from Emar.

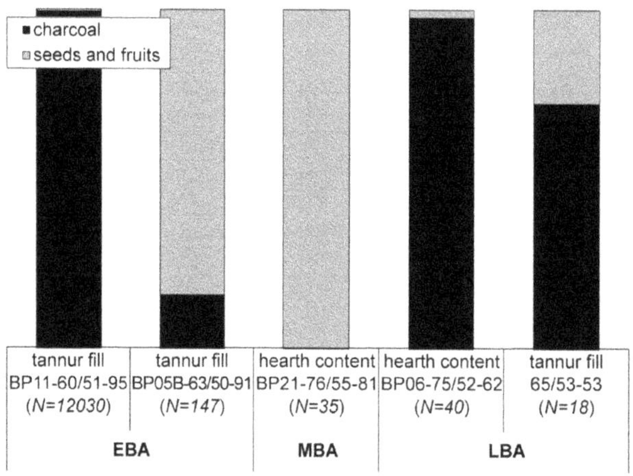

Fig. 10: Charcoal-to-seed/fruit proportions for oven (*tannur*) and hearth contexts from Emar.

There are several extant publications that allow us to gain further insight into the regional contemporary practises. Specific to the Upper Khabur, are two botanical studies that investigate the use of dung. The first details research at Leilan, about forty kilometres southeast of Mozan, which indicates the use of dung, especially within one Ninevite V (early third millennium BC) stratum. Their claim is backed by the high percentages of *Prosopis* weed seeds found there and which only occur long after the cereal crop harvest has been taken in from the fields.[39] Conversely, detailed investigations in to recognising dung-derived material at Tell Brak, located about forty kilometres south of Tell Mozan, did not indicate intensive use of dung as fuel; in fact, they showed that the Brak samples fit well in to a crop-processing model. However, dung pellets did occur within the charred botanical samples of Brak, albeit in less than 25% and the majority represented in the botanical samples consist of winter crop species.[40]

For Tell es-Sweyhat, thirty-two kilometres northeast of Emar, rather large proportions of seed-to-charcoal have been found in botanical remains that suggest the use of dung as fuel.[41] In comparison with Tell es-Sweyhat, preliminary research from Tell Jouwef, some three kilometres southwest of Tell-es-Sweyhat, reveals a somewhat larger proportion of charcoal while still containing signs of dung being used for fuel because significant quantities of dung were found within the samples. Furthermore, a more rigorous use of dung at Sweyhat can also be intimated by the ratio of wild seeds-to-cereals; with a much

[39] Wetterstrom 2003.

[40] Charles 1998.

[41] Miller 1997.

larger proportion of wild seeds retrieved at Sweyhat as compared to Tell Jouwef.[42] In spite of this, Jouwef is distinct in that typical floodplain woody vegetation was concurrently identified on-site as well, indicating riverine woodland was available within the reach of the Euphrates.[43] Sweyhat, on the other hand, is actually somewhat further removed from the Euphrates than Jouwef, which may be the explanation for its reduced access to riverine vegetation. Anthracological research in this area reveals that pistachio-almond woodland-steppe was probably present away from the river,[44] but it contained far less dense woody resources than the riverine forest.

Although the data sets are not easily compared, due to different methodologies applied to recognise dung as fuel, dung seems to have been more or less an additional source of fuel despite the presence of woodland in the neighbourhood of the sites mentioned. Therefore, dung was probably a fuel of choice and was perhaps used for specific purposes. For example, it may have particularly been used within certain types of ovens, as can be garnered from examples in Old Testament texts. These texts mention three different oven types used with different types of fuel: the bowl-oven, consisting of a large clay bowl that was inverted and placed upon small stones and heated with dried dung heaped over and around it; the large earthen-ware jar-oven was mentioned as being heated by fuels of grass, stubble, or dry twigs or thorns, and other things; and the pit-oven, which was formed partly in the ground and was partly built-up of clay and plastered throughout, narrowing towards the top, and fuelled by thorns and grass. According to the present data, no general rule can be identified that would indicate a different use of fuel within the ovens as compared with the hearths. Overall, the results appear to be rather varied, but the sample is too low to be statistically meaningful.

The employment of dung as fuel supports Wilkinson's interpretation of extensive off-site artefact scatters in the region under examination. Since dung was used as a fuel, it would have been unavailable or of reduced availability for application on the fields. The ash, which results from the burning of dung, was still valuable for crop nutrition and would have been deposited on fields. However, because this ash was more likely to derive from a settlement context, it was more likely to be deposited in conjunction with other inorganic settlement debris. Therefore, in regions where dung was intensively expended as fuel, extensive off-site scatters would be present.[45] Such typical artefact scatters have been found around some sites in the Upper Khabur Plain, such as at Tell Hamoukar, ninety kilometres east of Tell Mozan,[46] and within the Lake Assad region, at places like Tell Sweyhat.[47] Nevertheless, there seems to have been some local variations in its use. At Beydar, which also is located in the Upper Khabur Plains, no such evidence of an extensive artefact scatter away from the settlement has been found.[48] It has been put forward that manuring was not practised in this region and that the presence of a large area of pasture on the basalt plateau may have meant that dung was in plentiful supply, with wood fuel also available. In the latter case, dung may have been applied directly without much inorganic settlement-derived inclusions.[49]

6. Conclusion

Dung was probably used as fuel to a different extent at Tell Mozan, as compared to Emar. The qualitative charcoal data indicate woody resources were present at both sites, but the more intensive use of dung at Mozan (compared to Emar) can be explained by the riverine vegetation that occurred in the Emar surroundings; this riverine flora tends to be somewhat denser than the park woodland seen in reconstructions of the Mozan surroundings. However, it cannot be discounted that people used dung by preference, for example, for specific purposes. Finally, no special correlation was discovered between the use of dung as fuel and certain types of ovens or hearths.

The fact that dung was used as a fuel resource at Emar and Mozan is in line with archaeobotanical finds from other sites in the same region and also fits the presence of extensive artefact scatters around sites in this area that would indicate the use of burned dung manure. Nevertheless, dung was clearly not the only available and utilised fuel.

[42] Miller 2004.
[43] Miller 2004.
[44] Deckers – Pessin, chapter 3 this volume.
[45] Wilkinson 1989.
[46] Wilkinson 2003, 117-118.
[47] Wilkinson 2004, 55-72.
[48] Ur 2004.
[49] Ur 2004.

Bibliography

AKKERMANS, P.M.M.G. – SCHWARTZ, G.
2004 The Archaeology of Syria. From Complex Hunter-Gatherers to Early Urban Societies (c. 16000-300 BC).

ANDERSON, S. – ERTUG-YARAS, F.
1998 Fuel, fodder and faeces: an ethnographic and botanical study of dung fuel use in Central Anatolia, Environmental Archaeology 1, 99-109.

BARNARD, G. – KRISTOFERSON, L.
1985 Agricultural residues as fuel in the Third World, Energy Information Programme Technical Report 4.

BOTTEMA, S.
1984 The Composition of Modern Charred Seed Assemblages, in: W. van Zeist – W.A. Casparie (eds.), Plants and Ancient Man, 207-212.
1989 Notes on the Prehistoric Environment of the Syrian Djezireh, in: O.M.C. Haex – H.H. Curvers – P.M.M.G. Akkermans (eds.), To the Euphrates and Beyond. Archaeological Studies in Honour of Maurits N. van Loon, 77-87.

CHARLES, M.
1998 Fodder from dung: the recognition and interpretation of dung-derived plant material from archaeological sites, Environmental Archaeology 1, 111-122.

DECKERS, K. – RIEHL, S.
2007 An evaluation of botanical assemblages from the 3rd to 2nd millennium BC in northeastern Syria, Varia Anatolica XIX, 481-502.
2008 Tells of the Upper Khabur: their hydrological and agrarian contexts, Paléorient 34.2, 173-189.

ENGEL, T.
1993 Archaeobotanical analysis of timber and firewood used in the third millennium houses at Tall Bderi/Northeast Syria, in: K.R. Veenhof (ed.), Houses and households, 105-113.

GREGOR, H.-J.
1992 Paläobotanische Untersuchungen zur antiken Plfanzenwelt des Hassek Höyük im Tal des oberen Euphrat und ein Versuch zur Rekonstruktion der dortigen Ökosystems für den Übergang von der späten Urukzeit zur frühen Bronzezeit, in: M. Behm-Blancke (ed.), Hassek Höyük. Naturwissenschaftliche Untersuchungen und lithische Industrie, 34-67.

GÜNDEM, C.X. – UERPMANN, H.-P.
2003 Erste Beobachtungen an den Tierknochenfunden aus Emar (Syrien) – Grabungen bis 2002, Baghdader Mitteilungen 34, 119-128.

KING, L.W.
1896 Cuneiform Texts 1, British Museum.

KRAMER, C.
1980 Estimating prehistoric populations: an ethnoarchaeological approach, in: M.-T. Barrelet (ed.), L archéologie de l'Iraq: perspectives et limites de l´interprétation anthropologique des documents, 315-334.

LEMCKE, G. – STURM, M.
1997 $\delta^{18}O$ and trace element measurements as proxy for the reconstruction of climate changes at Lake Van (Turkey): preliminary results, in: N. Dalfes – G. Kukla – H. Weiss (eds.), Third Millennium BC Climate Change and Old World Collapse. Berlin, Springer, pp. 653-678.

MILLER, N.F.
1984 The use of dung as fuel: an ethnographic example and an archaeological application, Paléorient 10, 71-79.
1986 Seed-eaters of the ancient Near East: human or herbivore? Current Anthropology 37, 521-528.
1997 Sweyhat and Hajji Ibrahim: some archaeobotanical samples from the 1991 and 1993 seasons, in: R.L. Zettler (ed.), Subsistence and settlement in a marginal environment: Tell es-Sweyhat, 1989-1995, Preliminary Report, Museum Applied Science Center for Archaeology Research Papers in Science and Archaeology 14, 95-122.

2004 Flotation samples from the 1992 excavation at Tell Jouweif, in: T. Wilkinson (ed.), On the margins of the Euphrates. Settlement and land use at Tell es-Sweyhat and in the Upper Lake Assad area, Syria, 157-165.

MILLER, N.F. – SMART, T.L.
1984 Intentional burning of dung as fuel: A mechanism for the incorporation of charred seeds into the archaeological record, Journal of Ethnobiology 4, 15-28.

MOORE, A.M.T. – HILLMAN, G.C. – LEGGE, A.J.
2000 Village on the Euphrates.

PEÑA, L.P. – PEÑA-CHOCARRO, L. – ESTÉVEZ, J.J.I. – URQUIJO, J.E.G.
2003 Ethnoarchaeology in the Moroccan Jebala (Western Rif): wood and dung as fuel, in: K. Neumann – A. Butler – S. Kahlheber (eds.), Food, fuel and fields. Progress in African archaeobotany, Africa Praehistorica 15, 163-175.

PESSIN, H.
2004 Stratégies d'approvisionnement et utilisation du bois dans le Moyen Euphrate et la Damascène. Approche Antracologique comparative de sites historiques et préhistoriques, unpublished PhD dissertation.

PRUSS, A. – SALLABERGER, W.
2005 Tierhaltung in Nabada/Tell Beydar und die Bilderwelt der Terrakotten als Spiegel von Wirtschaft und Umwelt, Archiv für Orientforschung 50 (2003/04), 293–307.

RIEHL, S.
2010a The archaeobotany of Tell Mozan, in: K. Deckers – M. Doll – P. Pfälzner – S. Riehl (eds.), Ausgrabungen 1998-2001 in der Zentralen Oberstadt von Tall Mozan / Urkeš: the development of the environment, subsistence and settlement of the city of Urkeš and its region, SUN, Serie A, vol. 3.
2010b Maintenance of agricultural stability in a changing environment – the archaeobotanical evidence at Emar, in: U. Finkbeiner – F. Sakal (eds.), Emar after the closure of the Tabqa Dam. The Syrian – German Excavations 1996-2002, vol. I: Late Roman and medieval cemeteries and environmental studies, Subartu 25, 177-224.

UR, J.
2004 Urbanism and society in the third millennium Upper Khabur Basin, unpublished PhD dissertation.

VALAMOTI, S.M. – CHARLES, M.
2005 Distinguishing food from fodder through the study of charred plant remains: an experimental approach to dung-derived chaff, Vegetation History and Archaeobotany 14, 528-533.

WETTERSTROM, W.
2003 Ninevite 5 period agriculture at Tell Leilan: preliminary results, in: E. Rova – H. Weiss (eds.), The origins of North Mesopotamian civilization: Ninevite 5 chronology, economy, society, 387-400.

WICK, L. – LEMCKE, G. – STURM, M.
2003 Evidence of late glacial and Holocene climatic change and human impact in eastern Anatolia: high-resolution pollen, charcoal, isotopic and geochemical records from the laminated sediments of Lake Van, Turkey, The Holocene 13.5, 665-675.

WILKINSON, T.J.
1989 Extensive sherd scatters and land-use intensity: some recent results, Journal of Field Archaeology 16.1, 31-46.
2003 Archaeological landscapes of the Near East.
2004 On the margin of the Euphrates. Settlement and land use at Tell es-Sweyhat and in the Upper Lake Assad Area, Syria, excavations at Tell es-Sweyhat, Syria, vol. 1, Oriental Institute Publications 124.

ZOHARY, D. – HOPF, M.
2000 Domestication of plants in the Old World.

13. Wood use in the palace of Qatna

Katleen Deckers

1. Introduction

1.1 Archaeology of the site

Tall Mischrife is a large settlement hill in western Syria and has been identified as the ancient city of Qatna (fig. 1). It was initially inspected in 1924 and a second phase of investigations started in 1994 by a joint Syrian-Italian-German team.[1] The new excavations cover the periods from the Early Bronze Age (EBA) until the Iron Age II. The German excavations, from which the charcoal samples within this study derive, were conducted within the Royal Palace in area G.

The Royal Palace of Qatna was erected in the centre of the city on a natural cliff overlooking the northern Lower City. The Palace, which was the residence of the powerful rulers of the Kingdom of Qatna during the second millennium BC, was built in the Middle Bronze Age IIA (dated between 1800-1700/1650 BC) and remained in use until around 1340 BC (Late Bronze Age II), when it was destroyed at the time of the Hittite conquest of Syria.[2] It is one of the largest monumental palaces of ancient Syria with over eighty rooms on the ground level.

At the transition between the EBA and Middle Bronze Age (MBA), Mishrifeh´s urban layout changed drastically with, for instance, the building of the lower town fortification and construction of the palace.[3] Qatna, although situated at the margin of the Egyptian-controlled zone, seems to have been a loyal vassal of the Mitanni Empire during the Late Bronze Age (LBA). At about the mid fourteenth century, when the Hittite king Shuppiluliuma advanced along the Orontes, Qatna chose the protection of this new political force. After the Hittites took the city and destroyed the palace (ca. 1340 BC), it entered the sphere of influence of Pharaoh Amenhotep IV/Akhenaton. Textual evidence revealed that Qatna played a pivotal role in the trade between Mesopotamia and the Syrian coast.[4]

Underneath the palace foundations, EBA deposits have been excavated. Although they have not been extensively exposed, they seem to belong to domestic settlement contexts because the walls found indicate the presence of small rooms. In some places palace structures have been disturbed by Iron Age activities. Also noteworthy of mention, by about 1850 AD a village was constructed within this zone of excavations.[5]

1.2 Environmental setting of Qatna

Qatna is located within a region that receives nearly 300 mm of annual precipitation. The site also profits from a Mediterranean-type climate due to its geographic position east of the lowland. It is here that the Lebanon Mountains and the Syrian Jebel Ansariyah meet at roughly 34.5° N, allowing for a relatively far-reaching intrusion of the Mediterranean climate. Additionally, Qatna's position at the meeting point of three wadi tributaries from the Orontes provided the sustenance needed for settlement. In the past, Qatna was situated in an area with several subterranean karst springs but they have now been desiccated for roughly 20 years. Geomorphological research indicated the occurrence of lake deposits in connection with these karst springs. This lake within the river would probably have been created artificially by damming water near the site. Later on, at the start of the MBA, the lake would have been divided into two parts by the construction of the outer wall of Qatna; this means, a smaller portion was trapped inside the perimeter of the town and continued to be fed by a karst spring located at the northern foot of the acropolis. The largest part of the lake, however, constituted a sort of reservoir upstream for the city.[6]

---

1 Al-Maqdissi *et al.* 2002; Al-Maqdissi – Morandi Bonacossi 2005; Morandi Bonacossi 2007.

2 Pfälzner 2007.

3 Morandi Bonacossi 2007.

4 Klengel 2000.

5 Pfälzner 2007.

6 Cremaschi 2007.

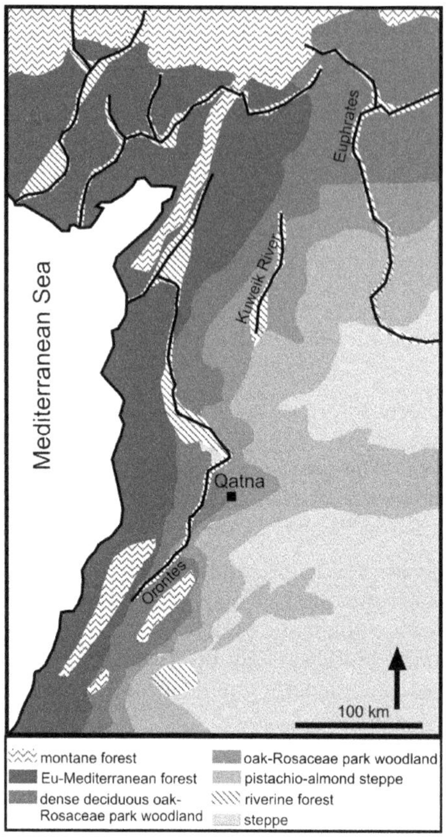

Fig. 1: Map with location of Qatna in its potential vegetation (according to Hillman in Moore *et al.* 2000).

Geobotanical assessments classify almost the entire Syrian territory into associations of the *Artemisietea herbae-albae mesopotamica* steppe.[7] Modern global ecosystem models place northern Syria in an area dominated by woody savannas, semi-desert shrubs, fields, and grass crops; also, most present-day layouts of potential vegetation show the area as dominated by shrubland, grassland steppe, and desert.[8] Recent anthracological research has indicated that park woodland and woodland steppe were more extensive in the past in northern and western Syria.[9] Potential vegetation reconstruction in Moore *et al.*[10] places Qatna in the borderland of a dense deciduous oak-Rosaceae woodland and an oak-Rosaceae park-woodland (fig. 1). In any case, this map is in greater agreement with basic textual information on the distribution of dense woodland in western Syria than are other maps of the potential natural vegetation.[11]

Until a short time ago, little was known about former vegetation in the Qatna surroundings. Recently however, palynological samples were retrieved from the Qatna surroundings and it was suggested that the MBA vegetation consisted of *Juniperus* (juniper) forest that also contained other trees, such as deciduous and evergreen oak.[12] In the course of the MBA, shortly before 1700 BC, *Juniperus* values declined and herbal pollen increased. This shift in the composition of the vegetation has been interpreted as an intensification of agricultural land-use and a dramatic transformation of the landscape into open land. Unfortunately, no pollen data is preserved for the LBA. Nevertheless, the trends visible during the MBA seem to have continued in the Iron Age.[13]

---

7 Zohary 1973.

8 NGDC 2006.

9 e.g. Deckers – Riehl 2007; Deckers *et al.* 2008; Deckers – Pessin, this volume chapter 3; Pessin 2004.

10 2000.

11 see discussion of historical vegetation in Wirth 1971.

12 Valsecchi 2007.

13 Valsecchi 2007.

1.3 Aims of this study

The overall aim of anthracological research at Qatna is to gain insight into the use of wood at the site and to reconstruct the vegetation of the surroundings. Compared to palynology, anthracology has the advantage that it can document the presence of insect-pollinated species (e.g. Rosaceae, maple) that will be underrepresented in pollen diagrams. Vegetation reconstructions using anthracology are mainly based on the assumption that charcoal samples from archaeological sites, especially charcoal deriving from fire wood, are representative of the availability of woody plants in the habitation environment. Therefore, the percentage of a charcoal taxon in the samples is used to indicate the relative abundance in the local vegetation.[14] However, human preferential selection may also have some influence on the charcoal fragment proportion of the site. Particularly within palatial contexts, it is not necessarily expected that local woody resources were exploited. Within this study special attention will be paid to the use of wood within palatial contexts.

2. Methods

Regular archaeobotanical sampling started in 2005 with the construction of a flotation machine and continued in 2006 under the direction of Simone Riehl. Aside from flotation samples, hand-picked samples were also retrieved. This is especially true for the royal tomb, where uncharred wood was also preserved; although it was often already heavily deteriorated and identification was therefore difficult. This is the case in a few uncharred samples from the tomb as well, where it was impossible to check diagnostic features to be certain of the preliminary identification. One such example of this is sample MSH07G-q1527, which is a conifer without resin ducts, possibly *Cedrus* (cedar) because of its omnipresence within the tomb, but the pitting on the tracheids could not be checked. Another problematic sample is MSH07G-q1313, a probable plane (*Platanus*) or beech (*Fagus*) fragment, but the longitudinal sections could not be investigated. Therefore, an attempt to differentiate between *Platanus* or *Fagus* based on the respective presence or absence of crystals in the rays could not be undertaken. It is more likely, however, that this sample is *Fagus* and not *Platanus* based on palynological research that has indicated *Fagus* occurred locally.[15] Additionally, poplar (*Populus*) and willow (*Salix*) cannot be easily distinguished from one another, and considering the poor preservation of the uncharred sample, no attempt was undertaken to differentiate between the two in this study.[16] Moreover, palynological research indicated that only willow was present locally.[17]

3. Details on sampled contexts

The location of all samples was mapped and catalogued in detail, so that their distribution could be plotted easily (figs. 2, 3, and 4).

Although the palace and the excavations cover a large area, archaeobotanical sampling that documents the use of wood within the palace could only be undertaken within the selected areas that are detailed herein. This is due to the fact that the palace had been previously excavated in the early twentieth century.[18] Moreover, later occupation remains disturbed the palace contexts. For example, the palace area was transformed into a sector of houses, domestic installations, storage pits, and handicraft installations in the Early Iron Age II (900-700 BC). Some of these contexts have been botanically sampled as well. But the fairly recent construction of a village led to the removal of even more palace wall remnants that had remained above ground, and were then made into new mud bricks for the villagers' houses.[19] Hence, at several locations only the palace foundations could be investigated to reconstruct the shape and function of the building. Charcoal remains within the MBA palace foundations could equally be the vestiges with a reused/mixed occupation from the EBA. Besides the EBA material thought to be reused within the MBA palace foundations, non-disturbed EBA contexts could have been sampled as well, but up until now only a few have been analysed. Since the excavation work among these deposits was relatively small, understanding of the contexts is somewhat difficult, but the finds seem to be located in a domestic sector.[20]

---

[14] see discussion in Asouti – Austin 2005, 2.

[15] Valsecchi 2007.

[16] Gale – Cutler 2000.

[17] Valsecchi 2007.

[18] Pfälzner 2007.

[19] Pfälzner 2007.

[20] Pfälzner 2007.

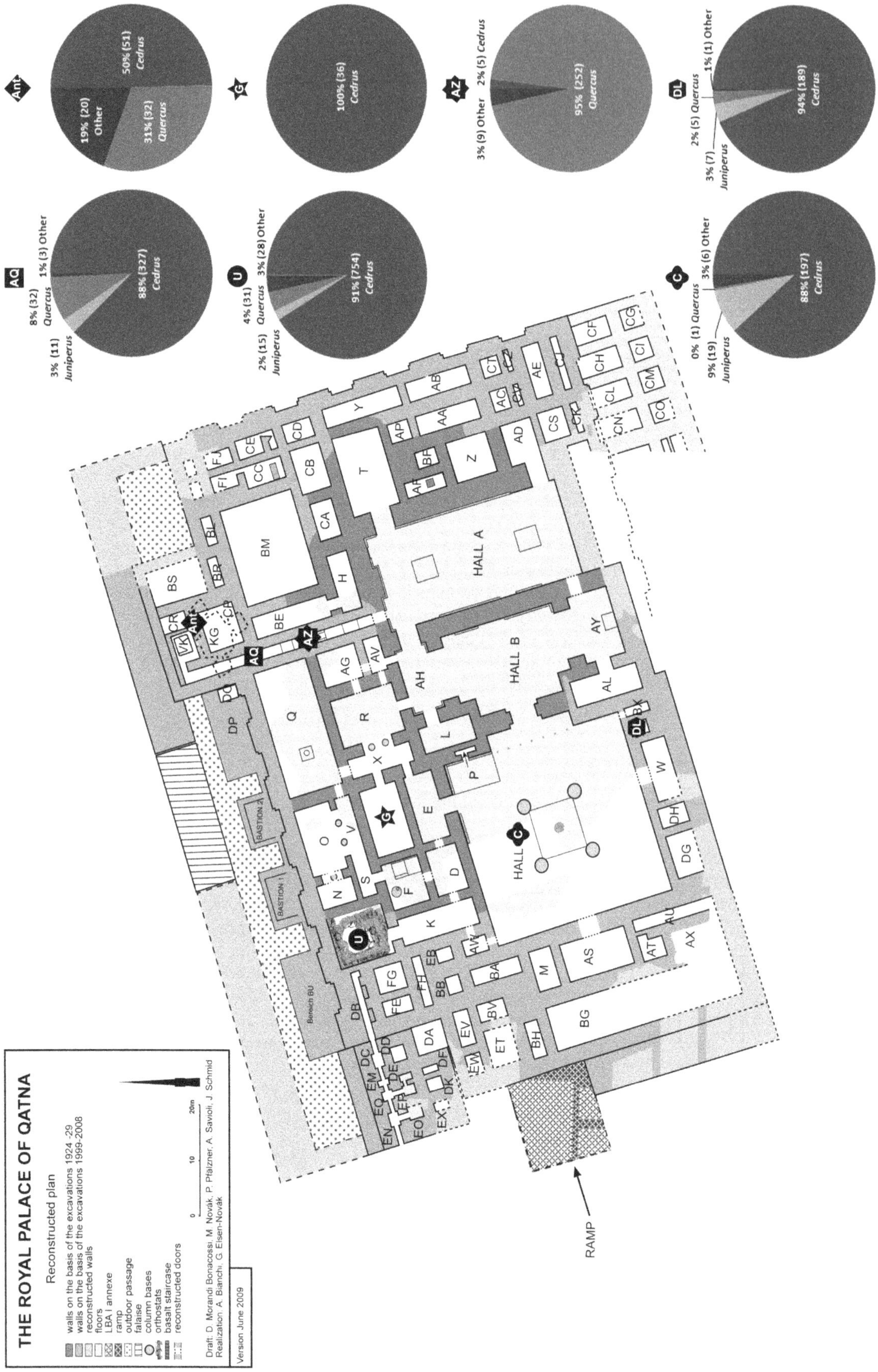

Fig. 2: Charcoal results plotted on the Qatna Palace plan.

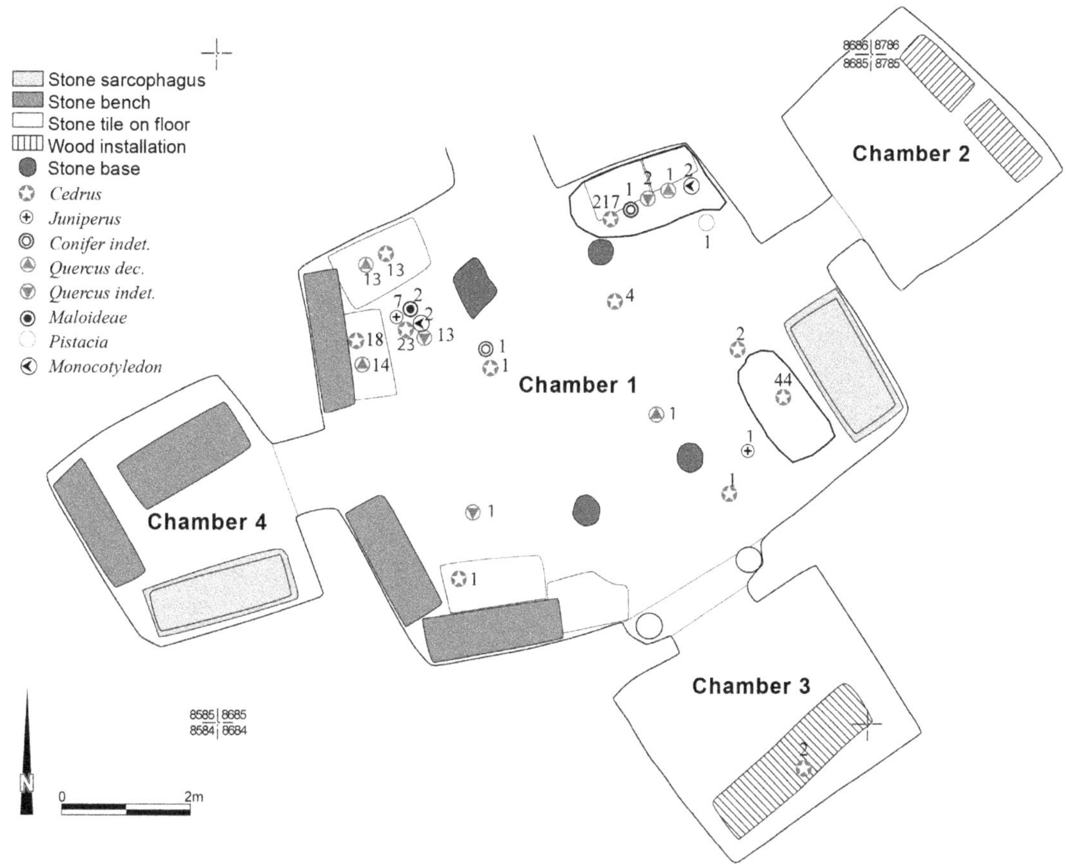

Fig. 3: Charred wood distribution within the Qatna tomb (numbers refer to numbers of fragments identified).

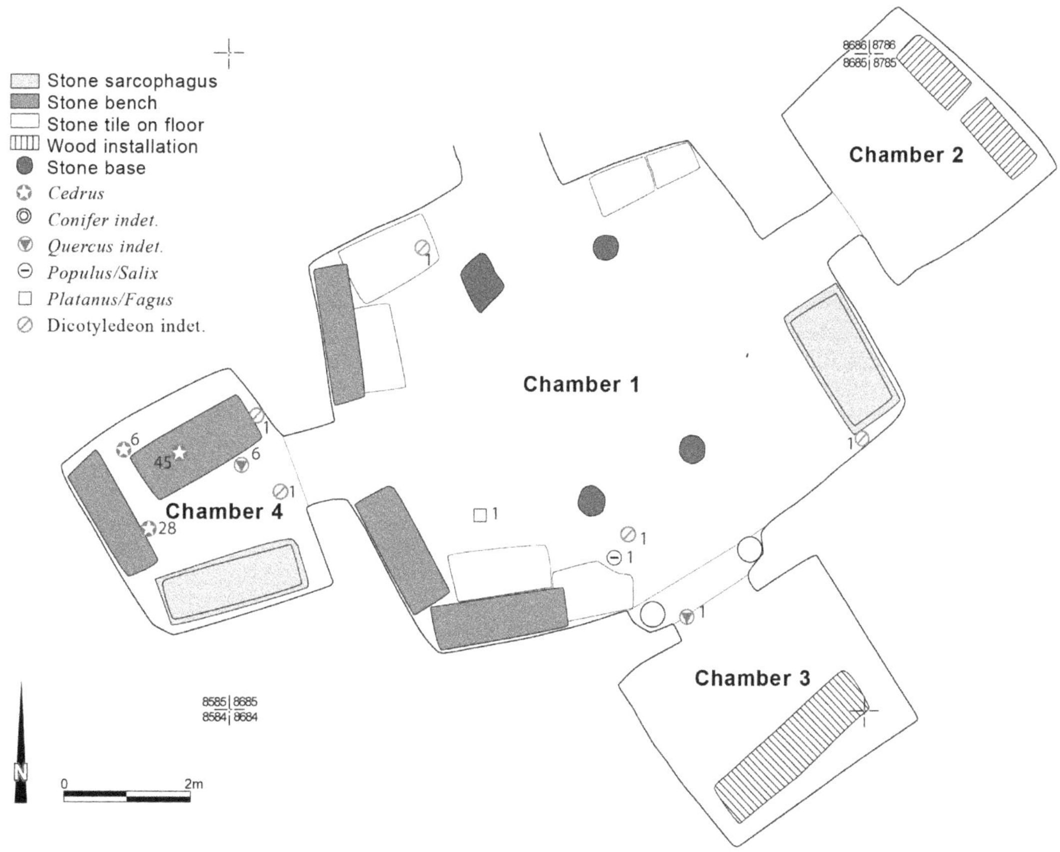

Fig. 4: Uncharred wood distribution within the Qatna tomb (numbers refer to numbers of fragments identified).

Only in exceptional locations did good sampling contexts for documenting wood use within the palace occur. For this reason, a judgmental flotation sampling strategy had to be adopted. Moreover, since the German mission is located only in the western part of the palace, charcoal samples could only be studied from this area; basically, only half of the palace was available for our study.[21]

Good botanical contexts that document the wood used within the palace occurred in the filled-up well, often referred to as the *cistern* (labeled as room U) (fig. 2). It is an impressive structure, measuring ten by ten metres with a depth of more than twenty metres and basalt steps leading downwards.[22] Within this structure, several metres of charcoal-rich sediments, sometimes even with whole beams, were deposited at once, probably upon the destruction of the palace. The beams and charcoal may derive from the roof or constructions above, or are due to adjacent rooms collapsing. One of the rooms adjacent to the cistern is where remains of wall paintings (room N) were found; it is interpreted to have had a cultic function. Another neighbouring room that was found was probably a bathroom (room F). To the west of the well, excavations indicate the presence of a kitchen area.[23] Hence, most of the charcoal samples from U represent architectural remains, although some contents of collapsed rooms may also have been included.

Another area where good botanical samples could be retrieved from the palace was within the eastern part of audience hall C (fig. 2). This enormous room, measuring thirty-six by thirty-six metres, had four large – now missing – wooden columns that supported the roof.[24] Sampled sediments for flotation within this room mainly represent a fire destruction collapse that was preserved only in parts of the room. Adjacent to hall C, a few samples derive from cellar DL in which more evidence for collapsing within the palace was also found.

A single non-flotation sample was retrieved from room G (fig. 2). This room appears to have been a special purpose storage room as it was not part of a large storage sector. Surprisingly, it still contained abundant remains of large-sized storage vessels when it was excavated for the second time, following the excavations of the 1920s.[25] The content of one of these vessels (MSH00G-i607) was sampled. This vessel was complete, though somewhat broken. Additionally, because its opening is relatively large, the vessel's contents do not necessarily only document its use but may also contain debris from a roof collapse.

Another area where palace contexts were sampled for botanical remains was within corridor AQ/AZ, which leads from the throne room (A) into the palace foundation (fig. 2). It contained a long staircase leading down with steps made of mud brick overlaid with wooden planks.[26] An exceptional feature within this corridor is the content of vessel MSH05G-i0083 which probably derives from the collapsed upper store. Due to this vessel's relatively narrow neck, it is possible that its contents do not relate to the destruction of the palace but are related to the use of the artefact itself.

At the end of the stairway, about 4 m below the palace floor, remains of a charred wooden door were found and botanically sampled (MSH02G-q547). Behind this door the corridor continues between narrow foundation walls. Here, debris from the fire on the ground floor of the palace had fallen. This corridor entered into the antechamber of the royal burial chamber. The antechamber, in turn, was filled with fire debris from the palace.[27] From the antechamber, 103 charcoal fragments from seven handpicked samples were investigated.

The tomb is the first in the region to have been discovered in an unlooted state in modern times. It is thought to have been in continuous use for 300-400 years for the burial of royal elite individuals.[28] All contents were found *in situ* and had never been buried nor come into significant contact with groundwater and thus were in a remarkable state of preservation. Almost 2000 finds were recorded, including jewellery, various carved objects, basalt statues, sarcophagi, human and animal bones, and roughly 200 pottery and stone vessels. Thirty-two charcoal samples and twelve wood samples from the chamber tombs were analysed, represented by 386 and ninety-three fragments respectively. Charcoal fragments were collected from locations throughout the tomb, but the highest concentrations were found within the northern reach of chamber 1 (fig. 3).

---

[21] The Italian mission also undertook extensive botanical sampling in the palace area, but their samples are still being studied.

[22] Pfälzner 2007.

[23] Pfälzner 2007.

[24] Pfälzner 2007.

[25] Pfälzner 2007.

[26] Pfälzner 2007.

[27] Pfälzner 2007.

[28] Pfälzner 2007.

The history of the charcoal within the tomb is somewhat unclear since no fireplaces or hearth installations were positively identified within the tomb. Although it is known that visits were regularly paid to the tomb and meals were offered to the corpses, it is concluded that meals were prepared outside the tomb, due to the absence of fireplaces and cooking pots.[29] It is likely that charcoal from the palace corridor ended up within the tomb upon the destruction of the palace. The excavators concluded that the concentration of charcoal remains near the entrance door of the chamber tomb was the result of the roof collapse of the rooms above the chamber tomb and antechamber. This was especially true of the charcoal concentrations within the northern reach of chamber 1 (fig. 1 and tab. 1).[30]

The theory that charcoal was thc result of the destruction of buildings from the palace does not hold for those locations further south within the chamber tombs. For example, a charcoal sample MSH07G-q1106 was found in chamber 3, on a wood installation that was probably a bed or bench.[31] Another similar area, where charcoals were found in association with a bed, is in the southeast of chamber 1 with samples MSH07G-q1274, MSH07G-q1249 and MSH07G-q1245. It can not be ruled out that the wood bed or benches themselves were charred because, based on the excavation database, the remains recovered would be uncharred wood installations themselves instead of the charcoal fragments lying upon the beds/benches.

A possibility as to how charcoal may have arrived in the chamber tombs is through the use of torches to light the tomb. This could especially be the case for charcoal found in chamber 2. These fragments have, however, been hitherto uninvestigated.

Uncharred wood was mainly preserved in chamber 4 but some pieces were also found at other locations. They often were attached to objects that were still located *in situ* so that it was possible to understand the original use of the wood.

4. Results

Charcoal identification results are listed in detail in Table 1 for the EBA contexts, in Table 2 for palace foundation contexts, in Table 3 for contexts that deliver information on the use of wood within the palace, and in Tables 4 and 5 for samples from the tomb of the kings.

Table 1: Charcoal results and context details of EBA Qatna samples.

EBA contexts	MSH05G-q455				MSH06G-q212	MSH05G-q351	MSH05G-q386
	BP21 (05)	BP17 (05)	BP18	BP20	BP 06/04 (06)	BP22 (05)	BP31 (05)
	Fill of room FD - location DL				Fill of gypsum Inst. 2186	Ashy fill - under hall C	Floor – under hall C
Cedrus				1	3	1	
Fraxinus							9
Maloideae							11
Monocotyledon						8	
Olea						9	
Pistacia atl.	1	4	1		5		3
Quercus dec.	1					1	
Quercus indet.					2		
indet.						2	2

Table 2: Charcoal results and context details of Qatna Palace foundation samples.

MBA palace foundation contexts	A8582	A8179				A8179				A3576	A8479
	G-q630	G-q411				G-q412				Unknown	G-q0482
	BP36 (06)	BP55 (05)	BP51 (05)	BP54 (05)	BP56 (05)	BP58 (05)	Unknown	BP57 (05)	BP59 (05)	BP62 (05)	BP22 (05)
	Foundation pit fill - Q	Foundation pit fill - hall C				Foundation - hall C				Foundation pit fill – hall C	Posthole content – hall C
Cedrus	1		1				3				
Fraxinus	67	1		10	2	1	1	3	3	6	
Monocotyledon	2	1									2
Olea	26			5	3						
Pinus halep.											5
Pistacia atl.		1						1		1	
Platanus/Fagus				1							
Populus/Salix	14	2		1	1				1	4	
Quercus dec.		2				4		1	11		
Quercus indet.							4		4		
Rosaceae spec.										1	
Ulmus			2						1		
Vitis	4	2		1	4					9	
indet.	3	1			2			1	4	2	

[29] Al-Maqdissi *et. al.* 2003.

[30] Pfälzner 2007.

[31] Witzel – Kreutz 2007, 178.

Table 3: Charcoal results and context details of palace samples Qatna (mainly use and collapse).

Context type	Fire destruction remains				Content of alabaster vessel MSH02G-i0277	Door remains	Filling from bowl	Fire destruction remains							Content of vessel MSH00G-i0607	Content of large basalt vessel MSH05G-i0083				Fire destruction remains											
Room	**C**	**C**	**DL**	**DL**	**AQ**	**AQ**	**AZ**	**Antechamber tomb**							**G**	**U**															
Cedrus	72	125	24	165	94	233	5	6			1		19	25	36	14	4	15	8	8	12	424	40	17	5	2	5	34	130	15	21
Chenopodiaceae																		1													
Ficus c.																			2												
Fraxinus		1							7	12												1			1		1				
Juniperus	13	6		7	11													1						6		6					2
Maloideae												1														1					
Monocotyledon			1																			12							1		
Olea					2	1										1															
Phragmites																													1		
Pistacia atl.																	1														
Quercus dec.			5		29		241							2		2	12	5	1		1	5					1				
Quercus evergreen									4																						
Quercus indet.		1			3		11		22		4								1						1		1				1
Vitis																						2									
indet.	3	2					9										1							1			1				

Table 4: Charcoal results and context details of samples from the Qatna tomb of the kings.

	In northeast corner of chamber 1															In northwest corner of chamber 1							In southeast corner of chamber 1				Centre of chamber 1			Chamber 3	Southwest of chamb. 1	South of chamber 1
	Under stone plate 1323					In between two parts of stone plate Inst. 1323		In between Inst. 1323 and wall	Eastern part of Inst. 1324	On floor	North of Inst. 1324	Besides stone plate Inst. 1324	On floor	On floor	Under stone plate Inst. 1330	Under stone plate 1333				Unknown	Under stone plate Inst. 1340		West of sarcophagus	West of sarcophagus	Under stone slab Inst. 2563	On floor	On floor	On floor	On floor	Inst. 1428 possible bed or table	Unknown	Possible wood floor cover
	MSH07G-q1503	MSH07G-q1501	MSH07G-q1505	MSH07G-q1505	MSH07G-q3164	MSH07G-q1497	MSH07G-q1500	MSH07G-q1624	MSH07G-q1686	MSH07G-q3210	MSH07G-q1267	MSH02G-q1176	MSH07G-q1628	MSH07G-q3220	MSH02G-q1126	MSH07G-q2339	MSH07G-q2347	MSH07G-q2344	MSH07G-q2358	MSH07G-q1226	MSH07G-q3171	MSH07G-q3174	MSH07G-q1274	MSH07G-q1249	MSH07G-q1619	MSH07G-q1245	MSH07G-q1151	MSH07G-q1148	MSH02G-q1200	MSH07G-q1106	MSH07G-q1237	MSH07G-q1260
Cedrus	8	5			26	9	4	1		30	77	3	8	47	4	1	2	14	1	23	11	2	42	1	2	1	1			2	1	1
Conifer indet.			1																									1				
Juniperus				1																7												
Maloideae					1															2												
Monocotyledon	1										1									2												
Pistacia (not *atlantica*)									1																							
Quercus dec.											1									13		13							1			
Quercus indet.	2																	11														

Table 5: Wood results and context details of samples from the Qatna tomb of the kings.

Context number	Location	Artefact type	Identification	No.
MSH07G-q1527	north of chamber 4	coffin	*Cedrus?*	28
MSH07G-q1223	north of chamber 4	coffin	*Cedrus*	45
MSH07G-q1539	northwest of chamber 4	coffin	*Cedrus*	6
MSH07G-q1530	entrance to chamber 4	golden cloak pin with wooden interior	indet. *dicotyledon*	1
MSH07G-q1307	southwest of chamber 1	wood installation	indet. *dicotyledon*	1
MSH07G-q1310	chamber 4, on floor east of bench	arrow	indet. *dicotyledon*	1
MSH07G-q1313	west of chamber 1	arrow	*Platanus/Fagus?*	1
MSH07G-q1312	southwest chamber 1	arrow	*Populus/Salix*	1
MSH07G-q1308	chamber 1, south of sarcophagus	bronze rivet with golden cap, including wood remains	indet. *dicotyledon*	1
MSH07G-q1540	chamber 4, south of bench	wood attached to bronze cramp	*Quercus?*	6
MSH07G-q1537	north of chamber 1	lance	indet. *dicotyledon*	1
MSH07G-q1311	entrance to room 3	bronze rivet with golden cap, including wood remains	*Quercus* indet.	1

Table 6: Summary of results from all Qatna samples by period and/or context type by number of fragments identified for each taxon, by fragment percentages for each taxon, by presence in number of samples for a certain taxon, and by ubiquity in percentages.

	Number of fragments					% of fragments					Ubiquity in samples					Ubiquity in %				
	palace contexts	tomb	EBA	palace foundations	Iron Age	palace contexts	tomb	EBA	palace foundations	Iron Age	palace contexts	tomb	EBA	palace foundations	Iron Age	palace contexts	tomb	EBA	palace foundations	Iron Age
Cedrus	1559	327	5	4	816	76.8	84.7	7.8	3.4	71.0	28	27	3	2	4	90.3	84	42.9	18	80
Chenopodiaceae	1				1					0.1	1				1	3.2				20
Conifer indet.		2					0.5					2					6.3			
Ficus c.	2					0.1					1					3.2		0.0		
Fraxinus	23		9	27	9	1.1		14.1	23.3	0.8	6		1	8	1	19.4		14.3	73	20
Juniperus	52	8			164	2.6	2.1			14.3	8	2			1	25.8	6.3			20
Maloideae	2	3	11	1		0.1	0.8	17.2	0.9		2	2	1	1		6.5	6.3	14.3	9.1	
Monocotyledon	14	4	8	3		0.7	1.0	12.5	2.6		3	3	1	2		9.7	9.4	14.3	18	
Olea	4		9	8	6	0.2		14.1	6.9	0.5	3		1	2	2	9.7		14.3	18	40
Phragmites	1					0.0					1					3.2				
Pinus halepensis				5					4.3					1					9.1	
Pistacia atl.	1		14	3	2	0.0		21.9	2.6	0.2	1		5	3	1	3.2		71.4	27	20
Pistacia not-atl.		1					0.3					1					3.1			
Platanus/Fagus				1	1				0.9	0.1				1	1				9.1	20
Populus/Salix				9	7				7.8	0.6				5	1				45	20
Quercus dec.	304	28	2	18	127	15.0	7.3	3.1	15.5	11.1	11	4	2	4	2	35.5	13	28.6	36	40
Quercus evergreen	4					0.2					1					3.2				
Quercus indet.	45	13	2	8		2.2	3.4	3.1	6.9		9	1	1	2		29.0	3.1	14.3	18	
Ulmus				3					2.6					2					18	
Vitis	2			16		0.1			13.8		1			4		3.2			36	
indet.	17		4	10	16	0.8		6.3	8.6	1.4	6		2	5	2	19.4		28.6	45	40
Total	**2031**	**386**	**64**	**116**	**1149**						**31**	**32**	**7**	**11**	**5**					
					3746															

Overall more than 18 taxa have been identified. The results are summarised in Table 6 according to context groups. At first glance, it is visible that within palatial and Iron Age contexts *Cedrus* is the main taxon represented, while the EBA and palace foundation samples only have small proportions of *Cedrus*. However, the palace foundation samples have a larger proportion and ubiquity of riverine vegetation (e.g. *Fraxinus, Platanus/Fagus, Populus/Salix,* and *Ulmus*) than the other samples. Additionally, the EBA and palace foundation samples contain larger proportions of fruit trees than those of other contexts.

Deciduous *Quercus* (oak) is represented in relatively large proportions and ubiquities in all Qatna samples. *Pistacia atlantica* (terebinth tree) is also present in almost all contexts, although in the EBA samples it has somewhat smaller proportions than *Quercus.* Finally, aside from Iron Age samples, Maloideae species also are present in most contexts and appear in equally small proportions as *Pistacia.*

5. Discussion

5.1 Early Bronze Age samples and local vegetation reconstruction

Since the EBA samples probably represent non-palatial contexts, they may reflect the local vegetation of the Qatna surroundings.[32] However, only a few samples have been investigated so far. Therefore, we must be careful not to over interpret them.

Pistacia atlantica was relatively evenly distributed throughout the samples. *Quercus*, some of which was deciduous whilst some could not be determined with certainty, also occurred regularly within the EBA samples. A Maloideae species was found in one sample and could have occurred in the same vegetation type as represented by the *Pistacia* and *Quercus* charcoal. Additionally, in one sample *Olea* charcoal was found that may represent a cultivated taxon.

The preliminary charcoal results support the potential vegetation reconstruction from Hillman,[33] placing Qatna in the oak-Rosaceae park-woodland. Up until now, no *Juniperus* has been found in EBA samples, although palynological research suggests it was the main component of the vegetation around Qatna in the MBA period.[34] Additionally, the find of *Olea* charcoal together with olive stones in EBA strata suggests that it was cultivated in the Qatna surroundings.[35]

[32] e.g. Asouti – Austin 2005.

[33] in Moore *et al.* 2000.

[34] Valsecchi 2007.

[35] Pena-Chocarro – Rottoli 2007.

5.2 Wood use in the architecture of the palace

The palace contexts and tomb entrance contexts that represent the caving in of the burnt roof at several locations are dominated by cedar, meaning it is very likely that the whole palace roof was covered mainly by cedar wood. However, *Juniperus* and *Quercus* are also frequently present in these contexts, though occurring in much smaller quantities. They may have also been part of the roof construction. For example, they might have been used as cross-tongues for the roof or were possibly part of the decoration.

If we assume that the whole palace was covered with a wooden roof, a surface of more than 14976 m^2 needed to be produced.[36] This would require quite a substantial amount of wood. A simple calculation may help us to visualize the amount of wood that was needed. If we assume that balks of around 10 × 0.33 × 0.33 m were chiefly used for the roof, about 5000 of them would be needed. For example, for every hectare of Cyprus forest, an average of eighty cubic metres of wood can be harvested, which equals about eighty trees.[37] This means that for the palace roof approximately 62.5 ha of forest would have been deforested.

Cedar wood was not locally available and may have been transported from the Lebanon Mountains, within a distance of approximately eighty kilometres, or from the north, with the nearest stands in the an-Nusariyeh Range in the Syrian coastal mountains about sixty kilometres away. These regions were under the control of the Qatna Kingdom during the MBA and LBA.[38] The transport of the wood to Qatna must have been very labour intensive.

5.3 Wooden artefacts and use of artefacts from the palace and the royal grave

In the royal graves, exceptional preservation circumstances allowed for the preservation of some wooden artefacts (tab. 5). Three arrowheads with attached wood remains were found. They are all made of different types of wood, one of which is probably beech (*Fagus*) and another willow. Willow is suitable for arrows and has often been used by Native Americans for this purpose. Beech wood is hard, strong and fine, with a regular and dense structure. It is easy to carve and bends well.[39]

A different object in chamber 4 has often been interpreted as a coffin and is made of *Cedrus*. This wood is aromatic and extremely resistant to insectival and fungal decay.[40] Therefore, it has often been used for coffins, as for example in the Midas Tomb of Gordion[41] or the coffin of Queen Merit Amon (fifteenth dynasty, fifteenth century BC) from Deir el Bahari.[42]

Additionally, it is of note that sample MSH07G-q1537, which may have belonged to a lance, seems to have been hardened by burning. Its outside was charred while the inside remained uncharred. This may support the theory that it was part of a weapon.

Another interesting find, is the fact that all charcoal fragments identified from the bed (MSH07G-q1274, MSH07G-q1249, MSH07G-q1619, MSH07G-q1245) were of the same taxon (*Cedrus*). This may indicate that they derive from one piece of furniture.

Charcoal remains within the doorway sample MSH02G-q547 of corridor AQ indicate that this door was also made of cedar, although a fragment of olive wood was found as well. This suggests that another context was dually sampled or that the door was made of several types of wood.

Another sample analysed (MSH05G-i83) was taken from a basalt bowl that was found located on the primary step installation of the well. Due to the very wide open nature of the bowl's mouth, it is uncertain whether a part of, all of, or none of the botanical samples retrieved represent remains related to its use. Compared to the botanical sample from MSH05G-i83, it is more likely that the one from within another stone vessel (MSH05-i277) contains remains related to its use; although the possibility that it contains debris from a palace roof collapse cannot be overlooked. However, it has been argued that this vessel had a purely decorative function because it has an open bottom, its ear (attached with bitumen) can not hold the weight of the vessel and it has a separately worked neck. The latter has a diameter of 6.3 cm.

Apart from the possible remains of a palace roof collapse (potentially represented by *Cedrus*, *Juniperus*, and *Quercus*), it is remarkable that both the wide bowl and the other vessel contained *Olea* wood remains.

---

[36] It has not been taken into account that there possibly was a second store.

[37] Constantinou 1982, 22-23.

[38] Pfälzner 2007, 35.

[39] Gale – Cutler 2000.

[40] Filley *et al.* 2001; Gale – Cutler 2000.

[41] Meiggs 1982; Filley *et al.* 2001.

[42] Donadoni 1969, 110.

Such wood remains did not occur regularly within the other contexts and therefore might uniquely be correlated to the use of these vessels.

It is likely that the botanical sample from a third vessel (MSH00G-i607) is also filled with roof collapse debris because it has a relatively large opening and contains *Cedrus*, a typical roof collapse content.

5.4 Secondary use of an earlier refuse dump in palace foundations

It is remarkable that the palace foundation's charcoal remains strongly resemble those of the EBA contexts and are different from the palace collapse and palace use contexts. This may indicate the secondary use of EBA refuse within the palace foundations. It also fits with thc ceramic identifications from the foundations, which reveal that all ceramic assemblages retrieved from original fills of foundation trenches throughout the palace contained a mixture of EBA and MBA materials.[43] However, as with the ceramics, there also may be MBA charcoals within the palace foundations.

5.5 Clues to the palace economy

The basic problem with functional and economic interpretations of the Royal Palace of Qatna is the fact that only foundations and floors survived, while walls, doors, installations, and inventories have largely been lost.[44] In spite of this, some hints of the palace economy and the use of space within the palace could be extracted from the fill of the well, the fill of corridor AQ, and room G. Another problem with reconstructing the palace economy from the botanical remains is that the area under investigation within this study does not seem to have included a large storage area. Such an area is located in the East of the palace, out of our research area.

LBA Levantine palace texts, such as those from Ugarit, often refer to vineyards and olive groves, indicating the importance of olive oil as well as grapes to the palace economy. The Ugarit texts from the fourteenth and thirteenth century BC mention that villages had to pay taxes to the palace in the form of oil and wine.[45] Based on text sources, Heltzer[46] calculated that the annual production of a village in the Ugarit neighbourhood was about 27.5 tons of oil. There existed about 200 villages in the Kingdom of Ugarit and it was estimated that the total oil production of the villages of Ugarit reached 5500 metric tons. Within the investigated palatial contexts at Qatna, olive wood is only represented in minor quantities. This correlates with the olive stone finds,[47] which were more numerous in domestic contexts.[48] Although olive wood is hard, durable, and strong, during the time period under consideration in the Levant, it probably came into the archaeological record mostly as the result of pruning, or through the use of unproductive or wild trees, and was used mainly as firewood. Olive wood can be burnt green, has a high calorific value and emits a pleasant odour.[49]

Grape (*Vitis*) was strongly represented in the palace area at Qatna by seeds[50] and, to a lesser degree, by wood remains; although the latter could be attributed to the palatial context, since *Vitis* wood does not generally attain great dimensions and therefore is limited in use. According to Klengel,[51] grapes for wine and olives for oil production were probably the most important economic plants in relation to supra-regional trade. In the texts from Ugarit, wine is mentioned more extensively than olive oil.[52]

As mentioned before, the cedar wood used in the palace is proof of considerable transport of wood from a distance of at least sixty kilometres, although probably still from within the territory of Qatna's control. Cedar was the typical resource that kings used to assert their power and which they demanded as tribute.[53] For example, in the third millennium BC Naram-Sin and Sar-kali-Sarri boasted of cutting Amanus cedars for temple constructions at Niniveh and Babylon,[54] and in the ninth century Shalmaneser III demanded up to more than 200 cedar logs annually from some of the Syrian princes who submitted to him.[55] Cedar wood

[43] Pfälzner 2007.
[44] Pfälzner 2007.
[45] Heltzer 1996.
[46] Heltzer 1996.
[47] personal communication with S. Riehl.
[48] compare with Pena-Chocarro – Rottoli 2007.
[49] Gale – Cutler 2000, 173.
[50] personal communication with S. Riehl.
[51] Klengel 2000, 246.
[52] Heltzer 1990.
[53] see e.g. Meiggs 1982.
[54] Frayne 1993, 140, 192.
[55] Meiggs 1982.

was typically used during this period in the Levant in areas where it did not naturally occur, such as the MBA palace of Lachish. Moreover, in the Bible (referring to the LBA and Iron Age) cedar wood is mainly mentioned when dealing with palaces and temples.[56] Hence, the cedar used within the palace of Qatna suggests that it was an exceptional procurement that represented the authority of the palace.

5.6 Indications for long-term vegetation change and environmental conditions

Although only a few samples that represent the local EBA vegetation have been analysed so far, the palace foundation's botanical samples are most likely mainly EBA vegetation remains as well. This is based on the fact that they probably represent the reuse of earlier deposits within the foundations.

Besides typical oak park woodland elements, riverine vegetation species have also been found within the botanical remains from the palace foundations and below. For example, *Fraxinus* (ash) had high ubiquity and fragment percentages in those charcoal samples. *Populus/Salix* charcoal (most likely *Salix* for reasons mentioned above) was also often present, though in relatively smaller numbers. Additionally, *Ulmus* (elm) occurred in a few samples. In the samples that are clearly from palace context these wood species are rarer or do not occur. *Fraxinus,* however, still occurs in palace destruction samples with a ubiquity of more than 15%. It is possible that a reduction of riverine vegetation was brought about by the intensive waterworks involved in the restoration of the city. More precisely, the establishment of the quadrangular, fortified earthworks caused the division of the previously undivided lake which was probably artificially created within the river,[57] and therefore caused ramifications to the riverine forest.

Juniperus, which according to the palynological evidence was strongly represented in the surrounding vegetation during the MBA, does not seem to have been used intensively in the EBA (at least not in the area under investigation). Therefore, the question arises as to whether it was strongly present during the EBA. It is known that *Juniperus* is a taxon that typically expands with human impact[58] and therefore may have been a secondary formation. The palynological samples indicate that shortly after 1680 cal. BC *Juniperus* rapidly decreased.[59] As we can conclude from the palace context charcoal samples, which regularly contained *Juniperus*, this reduction in pollen percentages may relate to the construction of the palace at about this time and/or may relate to long-term effects of the MBA settlement expansion that occurred somewhat earlier. Although the find of imported cedar remains within a palatial context is not surprising because it was a highly valued wood, it may also be symptomatic of a local decrease in woody resources.

Fruit tree charcoal taxa consisting of *Vitis* and *Olea* were strongly present in the EBA and in the palace foundation samples; the latter of which suggests their intensive cultivation in the surroundings during the EBA and MBA. Their reduction in the palace use contexts may relate to the fact that although hearths are found within palatial settings, ash from the hearths will normally be disposed of away from the palace. Moreover, *Vitis* wood does not generally attain large dimensions and is therefore even more limited in its use.

Finally, the presence of numerous traumatic resin canals in a large number of the *Cedrus* fragments might be indicative of some kind of environmental stress,[60] such as drought or frost. Nevertheless, traumatic resin canals can also be the result of human impact, like fire or the exploitation of resin by making shallow incisions on the tree. In some cases where traumatic resin canals were present, it could be observed that the annual rings were very dense, suggesting drought. However, this was not the case for all traumatic resin ducts. They were also present in fragments with less dense annual rings, indicating that they were not the result of drought.

6. Conclusion

The wood charcoal remains from the palace of Qatna indicate that large parts of the roof consisted of *Cedrus*, although it is possible that *Juniperus* and *Quercus* were used as well in the roof construction, albeit to a lesser extent. The large number of *Cedrus* charcoals found in several areas, and the overall surface area of the palace that would have been covered, attest to the enormous amount of *Cedrus* logs that must have been transported from relatively long distances. They manifest the power of the King of Qatna, but may dually intimate a depletion of woody resources in the local environment. In particular, the reduction in use

[56] Lipschitz 1992.

[57] Cremaschi 2007.

[58] Behre 1990.

[59] Valsecci 2007.

[60] Schweingruber *et al.* 2006.

of the riverine vegetation from EBA to palace use contexts (MBA or LBA) is remarkable and may relate to deforestation of this vegetation type. This deforestation was potentially related to the large water works that took place within the riverine environment from the MBA onwards and also may be tied to increased populations.

Although palatial texts indicate the importance of the role of fruit trees for the palace economy in the Levant, the Qatna Palace samples do not contain many charcoal remains of fruit trees. This may be due to the fact that garbage will normally not have been disposed of within the palace and that the contexts sampled have nothing to do with processing or storing of these crops. Nevertheless, it is remarkable that some of the vessels contained remains of olive charcoal; although we cannot rule out that they could have arrived there upon the destruction and collapse of the palace.

The wood and charcoal remains from the tomb of the kings provide information on the use of wood within the tomb. Some, such as the willow and beech which were used for arrows, were probably locally available. Others, like the wood for the cedar coffin, came from further away. The charcoal distribution from the tomb closely mirrors the percentages from the palace area, which may be partially explained by the collapsed rubble from the palace corridor falling in to the tomb upon its destruction.

Bibliography

AL-MAQDISSI M. – LUCIANI, M. – MORANDI BONACOSSI, D. – NOVÁK, M. - PFÄLZNER, P.
2002 Excavating Qatna. Preliminary report on the 1999 and 2000 Campaigns of the joint Syrian-Italian-German archaeological research project at Tell Mishrifeh.

AL-MAQDISSI, M. – DOHMANN-PFÄLZNER, H. – PFÄLZNER, P. – SULEIMAN, A.
2003 Das königliche Hypogäum von Qatna: Bericht über die syrisch-deutsche Ausgrabung im November-Dezember 2002, Mitteilungen der Deutschen Orient-Gesellschaft zu Berlin 135, 189-218.

AL-MAQDISSI, M. – MORANDI BONACOSSI, D.
2005 The metropolis of the Orontes. Art and archaeology from the ancient kingdom of Qatna. Seven years of Syrian-Italian collaboration at Mishrifeh/Qatna.

ASOUTI, E. – AUSTIN, P.
2005 Reconstructing woodland vegetation and its exploitation by past societies, based on the analysis and interpretation of archaeological wood charcoal macro-remains, Environmental Archaeology 10, 1-18.

BEHRE, K.-H.
1990 Some reflections on anthropogenic indicators and the record of prehistoric occupation phases in pollen diagrams from the Near East, in: S. Bottema – G. Entjes-Nieborg – W. van Zeist (eds.), Man´s Role in the Shaping of the Eastern Mediterranean Landscape, 219-230.

CONSTANTINOU, G.
1982 Geological Features and Ancient Exploitation of the Cupriferous Sulphide Orebodies of Cyprus, in: J.D. Muhly – R. Maddin – V. Karageorghis (eds.), Early Metallurgy in Cyprus, 4000-500 BC, Acta of the International Archaeological Symposium, Larnaca, Cyprus 1-6 June 1981, Nicosia, 13-24.

CREMASCHI, M.
2007 Qatna´s lake: a geoarchaeological study of the Bronze Age capital, in: D. Morandi Bonacossi (ed.), Urban and natural landscapes of an ancient Syrian capital; Settlement and environment at Tell Mishrefeh/Qatna and in Central-Western Syria, Studi Archeologici su Qatna, Forum: Udine, 93-104.

DONADONI, S.
1969 Treasures of the Egyptian Museum, Cairo.

DECKERS, K. – RIEHL, S.
2007 An evaluation of botanical assemblages from the 3rd to 2nd millennium BC in northeastern Syria, Varia Anatolica XIX, 481-502.

DECKERS, K. – RIEHL, S. – JENKINS, E. – ROSEN, A. – DODONOV, A. – SIMAKOVA, A.N. – CONARD, N.J.
2008 Vegetation development and human occupation in the Damascus Region of southwestern Syria from the Late Pleistocene to Holocene, Vegetation History and Archaeobotany 18.4, 297-314.

FILLEY, T.R. – BLANCHETTE, R.A. – SIMPSON, E. – FOGEL, M.L.
2001 Nitrogen cycling by wood decomposing soft-rot fungi in the "King Midas tomb," Gordion, Turkey, Proceedings of the Natural Academy of Sciences of the United States of America 98 (23), 13346-13350.

FRAYNE, D.
1993 Sargonic and Gutian Periods (2334-2113 B.C.), The Royal Inscriptions of Mesopotamia Early Periods 2.

GALE, R. – CUTLER, D.
2000 Plants in archaeology. Identification manual of vegetative plant materials used in Europe and the southern Mediterranean to c. 1500.

HELTZER, M.
1990 Vineyards and Wine in Ugarit (Property and Distribution), Ugarit-Forschungen 22, 119-135.
1996 Olive growing and olive oil in Ugarit, in: D. Eitam – M. Heltzer (eds.), Olive oil in Antiquity, Israel and Neighbouring Countries, 77-90.

KLENGEL, H.
2000 Qatna - ein historischer Überblick, Mitteilungen der Deutschen Orient-Gesellschaft zu Berlin 132, 239-252.

LIPSCHITZ, N.
1992 Building in Israel throughout the ages – one cause for the destruction of the cedar forests of the Near East, GeoJournal 27.4, 345-352.

MEIGGS, R.
1982 Trees and timber in the ancient Mediterranean world.

MOORE, A.M.T. – HILLMAN, G.C. – LEGGE, A.J.
2000 Village on the Euphrates: from foraging to farming at Abu Hureyra.

MORANDI BONACOSSI, D.
2007 Qatna and its hinterland during the Bronze and Iron Ages. A preliminary reconstruction of urbanism and settlement in the Mishrifeh Region, in: D. Morandi Bonacossi (ed.), Urban and natural landscapes of an ancient Syrian capital. Settlement and environment at Tell Mishrefeh/Qatna and in Central-Western Syria, Studi Archeologici su Qatna, 65-92.

PENA-CHOCARRO, L. – ROTTOLI, M.
2007 Crop husbandry practices during the Bronze and Iron Ages in Tell Mishrifeh (Central-Western Syria), in: D. Morandi Bonacossi (ed.), Urban and natural landscapes of an ancient Syrian capital. Settlement and environment at Tell Mishrifeh/Qatna and in Central-Western Syria, Studi Archeologici su Qatna, Forum: Udine, 123-145.

PESSIN, H.
2004 Stratégies d´approvisionnement et utilisation du bois dans le Moyen Euphrate et la Damascène. Approche Antracologique comparative de sites historiques et préhistoriques, unpublished PhD dissertation.

PFÄLZNER, P.
2007 Archaeological Investigations in the Royal Palace of Qatna, in: D. Morandi Bonacossi (ed.), Urban and natural landscapes of an ancient Syrian capital. Settlement and environment at Tell Mishrifeh/Qatna and in Central-Western Syria, Studi Archeologici su Qatna, Forum: Udine, 29-64.

SCHWEINGRUBER, F.H. – BORNER, A. – SCHULZE, E.-D.
2006 Atlas of woody plant stems: evolution, structure and environmental modifications.

VALSECCHI, V.
2007 Vegetation and environmental changes during the Middle-Late Holocene at Tell Mishrifeh/Qatna: climate versus land-use, in: D. Morandi Bonacossi (ed.), Urban and natural landscapes of an ancient Syrian capital. Settlement and environment at Tell Mishrifeh/Qatna and in Central-Western Syria, Studi Archeologici su Qatna, Forum: Udine, 105-114.

WIRTH, E.
1971 Syrien: Eine geographische Landeskunde.

WITZEL, C. – KREUTZ, K.
2007 First results of the anthropological and palaeopathological examination of the human skeletal remains recovered from the Royal Tomb of Tell Mishrifeh/Qatna, in: D. Morandi Bonacossi (ed.), Urban and natural landscapes of an ancient Syrian capital. Settlement and environment at Tell Mishrifeh/Qatna and in Central-Western Syria, Studi Archeologici su Qatna, Forum: Udine, 173-188.

ZOHARY, M.
1973 Geobotanical Foundations of the Middle East, vol. 2.

Lightning Source UK Ltd.
Milton Keynes UK
UKHW052157140120
356959UK00003B/34/P

9 782503 541068